THE COMPLETE BOOK OF CHEVROLET

CAMARO®

SECOND EDITION

EVERY MODEL SINCE 1967

DAVID NEWHARDT

motorbooks

Brimming with creative inspiration, how-to projects, and useful information to enrich your everyday life, Quarto Knows is a favorite destination for those pursuing their interests and passions. Visit our site and dig deeper with our books into your area of interest: Quarto Creates, Quarto Cooks, Quarto Homes, Quarto Lives, Quarto Drives, Quarto Explores, Quarto Gifts, or Quarto Kids.

First published in 2012 by Motorbooks, an imprint of The Quarto Group, 401 Second Avenue North, Suite 310, Minneapolis, MN 55401 USA.
T: (612) 344-8100 F: (612) 344-8692
www.QuartoKnows.com

Motorbooks titles are also available at discounts in bulk quantity for industrial or sales-promotional use. For details contact the Special Sales Manager at The Quarto Group, 401 Second Avenue North, Suite 310, Minneapolis, MN 55401 USA.

ISBN: 978-0-7603-5336-3

ACQUIRING EDITORS: Chris Endres and Darwin Holmstrom
PROJECT MANAGER: Jeff Zuehlke and Jordan Wiklund
COVER DESIGN: Cindy Samargia Laun
BOOK DESIGN: Kou Lor and Simon Larkin
LAYOUT: Chris Fayers and Rebecca Pagel

ENDPAPERS: A 2017 Chevrolet Camaro burnout (front) and fifty years of "conquering cousins" (back). *GM*
FRONT COVER: The 2017 Chevrolet Camaro 1LE package. *GM*
BACK COVER: 2017 Chevrolet Camaro ZL1. *GM*
TITLE PAGE: 2017 Camaro interior. *GM*

Printed in China

CONTENTS

ACKNOWLEDGMENTS

Like so many endeavors that give a single individual credit, the truth is different than first impressions. Writing and photographing a book is a team effort in the truest sense of the word. Vehicle owners, interviews, patient editors; these are the people which make a book like this become a reality. I would like to list these wonderful enthusiasts, in no particular order. To all of them, I extend a very heartfelt "Thank You"!

Dale Berger Jr., Matt Berger, Stefano Bimbi, Peyton Cramer, Joel Rosen, Marty Schorr, Charley Lillard, Mike Guarise, Lynn Shelton-Zoiopoulos, Dana Mecum, Jennifer Scott-Roshala, Louise Bent, Bill & Wanda Goldberg, David Christenholz, Mark Hassett, Colin Comer, Tammy & Gib Loudon, Rodney W. Green, Darwin Holmstrom, my editor Jeff Zuelhke, and publisher Zack Miller.

The 1979 Z28 can be distinguished from the 1978 version by the new front three-piece spoiler and front flairs.

INTRODUCTION

The times they were a-changin'. A new generation, born in the immediate postwar years, was now entering society with money in their pockets and firm ideas about what they wanted and what they didn't want. And one of the things they didn't want was a car that looked like what Mom and Dad drove. To these baby boomers, their parents had driven bulbous, hulking machines that looked old the day they were built. Station wagons, boring sedans . . . no, the boomers wanted excitement on wheels, a mechanical partner that was young looking, fleet-footed, and inexpensive.

Ford Motor Company read the tea leaves well and gambled on a new vehicle built on the bones of a sturdy economy car. Ford felt that the new Mustang was a secretary's car and didn't have illusions that it would set the motoring world on fire. But that's exactly what the "pony car" did. Dearborn was overwhelmed by the public's reaction to this simple and cheap 2 + 2 equipped with either a six-cylinder or a small-block V-8.

It's said that imitation is the sincerest form of flattery, and General Motors wasn't above stealing a page from another's playbook if it meant that it could "borrow" the concept to create a similar vehicle to the Mustang. And that's exactly what GM did. By using the Nova economy car as the basis for a long-hood/short-deck sporty car, GM could price it competitively to the Mustang. A vehicle's chassis/platform is traditionally the most expensive and time-consuming part to develop, and by utilizing the Nova's capable if boring underpinnings, Chevrolet could field a competitor to the dreaded Mustang in a decent amount of time. And that's exactly what it did.

The response was heartening to Chevrolet, who knew that there was a body of buyers who wouldn't get caught dead in a Ford, but who were also frustrated that GM in general, and Chevrolet in particular had not manufactured an affordable competitor. These folks flocked to Chevrolet dealerships, and soon a genuine battle for a narrow slice of the market was in full fight. The contest would last for decades, with the lead swinging in both directions. As the twenty-first century got rolling, the plug was pulled on the Camaro, but soon cooler heads prevailed, and within a few years showrooms were once again stocked with new Camaros. The struggle for this slice of the sales pie continues to this day, with the consumer being the lucky recipient of Chevrolet's efforts. What will the Camaro of the future look like? One thing is for sure, it will cause hearts, young and old, to skip when one rolls by. While the youth of the 1960s wanted a fun car to show off in, today all ages want to show off in a Camaro.

1

GENERATION ONE

Camaro enthusiasts turn toward Dearborn, Michigan, and give a nod of thanks to General Motors' arch nemesis, the Ford Motor Company. The reason is clear: Without the Ford Mustang to appeal to the baby boomer generation, the Camaro would not exist. It took the staggering sales numbers that the Mustang generated, right from its inception, to wake the complacent General Motors up to a huge marketing niche that it was ignoring.

Lido "Lee" Iacocca was the General Manager at Ford Division, and he had a finger on the pulse of the youth market. He formed the Fairlane Committee, a group assembled to brainstorm new products. One of the first vehicles from the committee was the Falcon Futura, a huge-selling economy car. The committee's next idea was a car built on the Falcon platform.

Ford Motor Company, dwarfed by rival GM, had to use as much off-the-shelf technology as possible to afford a "new" vehicle. The economy-based Falcon had sold enough units over the years to amortize the platform costs. It was cheap, and it could handle six-cylinder and small eight-cylinder engines,

In many people's mind, a sports car isn't a sports car if there is a roof over their heads. Chevrolet knew this, and when the Camaro debuted in late 1966, each version of the car except for the Z/28's 302 V-8 was available in a ragtop Camaro.

9

This unassuming 1964½ Mustang coupe was the first coupe sold and the second production Mustang built. General Motors was stunned when this car, based on an economy platform, broke sales records. At the time, GM had no effective response, sending the corporation into a panic.

The rear-engine, air-cooled Corvair was, in the early 1960s, Chevrolet's hope against the Ford Mustang, but it didn't take long to realize that the Corvair wasn't going to put a dent in Mustang sales. Drastic action had to be taken, hence the birth of the Camaro. *Author collection*

important when marketing to a wide range of potential buyers. The newest vehicle's bodywork was a departure from traditional Detroit offerings, with their huge trunks, roomy interiors, and imposing front-end sheet metal. The car, called the Mustang, moved the passenger area rearward, shrinking the back seats, and giving the trunk a case of miniaturization. The result was a revolution, and a genre called the pony car was born.

Sporting a long-hood/short-deck profile, the 1964½ Mustang (named after the World War II American fighter aircraft) looked like nothing else coming out of Motor City. Its sporty lines spoke to the young and the young at heart; and while it was originally intended to be a "secretary's car," enough were equipped with healthy V-8 engines to give the Mustang some serious street credibility. The public ate it up, to the tune of 418,812 sold at the end of the 1965 model year. That number got GM's attention.

But some in General Motors felt that they had a viable competitor to the Mustang, a car that had been in production since 1960: the Corvair! This diminutive, rear-engine, air-cooled passenger car was sporty (sort of), but its swing-axle rear suspension design and the hungry ambition of a young attorney resulted in a book, *Unsafe at Any Speed*,

Above: A Chevrolet engineer poses with the six-cylinder, opposed-piston, air-cooled aluminum engine fitted to the Corvair. This reliable powerplant was mounted in the rear of the Corvair, and while it competed with other economy cars, it was not, and never would be, a muscle car. *Author collection*

Left: The 1967 Camaro debuted with two body styles, a coupe and convertible. This period Chevrolet image used a billboard to help show both bodies in a single shot. Waving at billboards is an activity not seen anymore. *Author collection*

Trim, conservatively styled, well-priced, and a bit boring, the Corvair
in 1963 still suffered from a rear suspension that lawyer Ralph Nader
took to task in his book *Unsafe at Any Speed*, dooming the Corvair.
Even though its rear suspension was re-engineered for 1965, the
damage was done, and the Corvair was shown off the stage.

which essentially killed off the Corvair. Even in its best year, the Corvair never posted sales figures that kept any Ford executives awake at night. But Chevrolet did have the Chevy II Nova, and for 1965 the Corvair was restyled and re-engineered resulting in a safe, fun-to-drive automobile. But the Corvair's fate was sealed, and Chevrolet knew that it had to field an entry into the pony car arena quickly.

Fortunately, General Motors had some executives that understood cars, and they had seen the signs of an impending youth market. GM's design studio, led by the famed Design Vice President Bill Mitchell, had released the stunning Riviera in late 1962 as a foil to Ford's Thunderbird. Irv Rybicki, the group chief designer at Chevrolet, liked what he saw with

the Riviera and pushed for a Chevy version of a sport-coupe. Mitchell backed Rybicki up, and quietly, plans were laid for a Bowtie personal car.

Rybicki and his team didn't have a clue that Ford was developing the Mustang on an economy car platform, but that's the same path that the Chevrolet designers followed, using the Chevy II as the donor. Chevrolet Division Manager Bunkie Knudsen liked what he saw, but he had another new car in the pipeline: the Chevelle. While the Chevelle was to be built on an intermediate A-body chassis, Knudsen was reluctant to put yet another vehicle into the market, feeling that might siphon off sales of other Chevrolet products. So the plan for a mini-Riviera was put on hold.

Bucket seats do not make a sports car. While some General Motors decision makers felt that the Corvair was a valid competitor to the 1964 Ford Mustang, the massive sales numbers generated by the Mustang soon caused these same executives to realize that they needed an effective front-engine, rear-drive sporty car, pronto!

The new Camaro would adopt the Nova's hybrid chassis, which featured unit construction for the main body but used a traditional rail-type subframe under the engine compartment. *General Motors 2012*

Fortunately, over in Chevrolet's Design Studio 2 in 1964, its chief, Hank Haga, had his stylists create a vehicle they called the Super Nova, based on the Chevy II. It was clad in a fiberglass body, painted Fire Frost silver, and shown at the 1964 New York Auto Show. Looking not a little like a Chevy version of a mini-Riviera, it was a sleek vehicle built on a Chevy II's 110-inch wheelbase platform, loaded with gee-whiz features that didn't stand a chance of production at the time, such as an Elometer, an elapsed time meter, and solenoid activated door releases. The crowds at the auto show liked the Super Nova, but Chevrolet hemmed and

hawed about putting it into production. As it sat beneath the bright lights at the auto show, Ford executives circled the Super Nova like vultures over a fresh kill; they had plans of their own and the Super Nova had hit a nerve. Most of the decision makers at Chevrolet wanted to put some form of the Super Nova into production. Bunkie Knudsen pondered how he could get his bosses to green light the vehicle. But General Motors President Jack Gordon decided to kill it. Chevrolet was left with the Corvair to hold down the small, fun-to-drive segment. But four months later, a bomb landed on General Motors, shaped like a Ford Mustang.

Above: While handsome enough in its own right, the staid lines of the Chevy II would not suffice for a sporty Mustang competitor. *General Motors 2012*

Left: Chevrolet needed to move fast to compete with Ford's Mustang, so it based its new pony car on the Chevy II platform, which was GM's most advanced small car at that time. *General Motors 2012*

The 1964 Chevrolet Super Nova concept car was constructed on the Chevy II economy car. But
with its sleek lines and 7.5-inch lengthening of the nose, it was a graceful experiment. A number of
stylistic elements were later used on the 1967 Camaro, including the hood vents and a mild V-grille.
With a body made of fiberglass, the 1964 Super Nova looked a bit like a baby Riviera, but with its
Chevy II underpinnings, including the 110-inch wheelbase, it showed what Chevrolet's designers
were considering to combat the Mustang.

INDEPENDENT FENDER BLISTERS

FRONT EMPHASIS

Knudsen spoke the corporate line that the Corvair was just what was needed to combat the Mustang scourge. But as Mustang sales climbed to unheard of levels, such as 100,000 units in just four months, Knudsen realized that some GM executives might have misread the market and put the pressure on to come up with a suitable vehicle, pronto. Bill Mitchell was ready to develop the concepts that Haga and Rybicki had created, and Mitchell had the clout to make things happen quickly.

A number of prerequisites were clear from the beginning. The new vehicle had to be superior to the Ford Mustang in *every* category. But by the time Chevy's sporty car would hit the showroom, the Mustang name would have over two years to burrow into the buying public's subconscious. Chevrolet knew that parity with the Mustang wouldn't cut

it; the new car had to deliver a crushing blow in the form of a vastly superior vehicle, with a superb ride, strong powerplants, tight and even build quality, nimble handling, and head-turning looks. No pressure at General Motors. The new vehicle, internally coded XP-836, then "F-body," had to best the Mustang without reservation. General Motors knew that the Mustang had a significant head start in the market, and when the Chevrolet vehicle, code-named Panther, finally hit the market, the second-generation Mustang would be rolling out. In order to get the car in showrooms as quickly as possible, it was necessary to use as much existing General Motors material as possible to minimize designing new parts. Chevrolet's Chevy II economy car was tapped as the donor platform for the new Mustang-fighter, exactly the same path Ford had taken to develop the Mustang by using the Falcon.

From the earliest clay models, it was clear that the Camaro would have cutting-edge styling. Some of the clay models produced early in the development of Chevrolet's F-body car had far more exotic styling than would actually appear on the production car. *General Motors 2012*

Above: The XP-836 design exercise was much closer to the production Camaro than earlier clay models. *General Motors 2012*

Right: The lines of the XP-836 might have been a bit swoopier than those that would appear on the production Camaro, but overall, the production car adhered closely to the model's lines. *General Motors 2012*

One challenge was already clear to the Chevrolet engineers: the Chevy II, in stock form, had all the sporty verve of a wheelbarrow. Originally designed as basic transportation, the Chevy II had been built to a price-point. In order to make the Chevy II as fiscally attractive as possible, little was done, especially engineering-wise, in an effort to create a taut road machine. Nobody bought a Chevy II to make a performance statement. As long as the occupants were kept dry and warm on a rainy night, the Chevy II fulfilled its purpose. Now Chevy's engineers were tasked with making a silk purse out of a sow's ear, and doing it as quickly as possible. Nothing like a challenge, but General Motors had some of the best engineers in the world. And they had a new tool that was just entering their world: a computer. Not exactly a supercomputer by today's standards,

its punch cards were used to check design work on the body and chassis, as well as develop prototype parts. Anything to speed the process. Chevrolet was in a race, and it had started from far, far back in the pack.

In order to cut costs and weight, the Chevy II had been designed as a semi unit-body vehicle. From the firewall back, the Chevy II did not have separate body-on-frame construction; the body was the frame. But from the firewall, or cowl, forward, a stub frame was bolted onto the body shell. On the new Camaro, the arms of the subframe would extend beneath the body to the area under the front seats. This subframe held the engine, transmission, brakes, inner and outer sheet metal, and front suspension and was attached to the body with rubber "doughnuts" surrounding the bolts in an attempt to minimize the transmission of noise, vibration,

When the 1967 Camaro debuted, the base engine was a straight-six cylinder mill. Many wanted a few more beans—and cylinders—under the hood, and the 350-cubic-inch V-8 was a durable and strong replacement. It weighed just 72 pounds more than the six-banger, and in SS configuration, it delivered a stout 295 horsepower.

Above: While the Camaro featured no groundbreaking engineering, it did represent the state of the automotive art when it appeared in the fall of 1966. *General Motors 2012*

and harshness (NVH) into the interior. That's a good plan, if you don't mind a degree of flex between the subframe and body. This flex gave the Chevy II some unique handling characteristics, none of which engendered confidence on the road.

To improve the ride and handling of the new car beyond the Chevy II, Chevy engineers extended the length of the subframe portion that fit underneath the body shell. This increased contact area helped stabilize the subframe in relation to the body. Four tuned rubber mounts were used to firm up the ride and improve steering accuracy. The result of this work was

successful, giving the F-body a far superior ride to the donor Chevy II and a more refined feel than the Mustang.

The front suspension was given considerable attention to bring it to an acceptably sporty level. Featuring short upper control arms and a long wishbone-style lower control arm configuration, it used coil springs with shock absorbers inside, as well as a crossmember, to link the entire front suspension together. To fight body roll, all F-bodies were fitted with a 0.687-inch anti-sway bar in the bow. The steering gear was a parallel relay recirculating-ball setup mounted behind the suspension. At the other end of the car, the solid

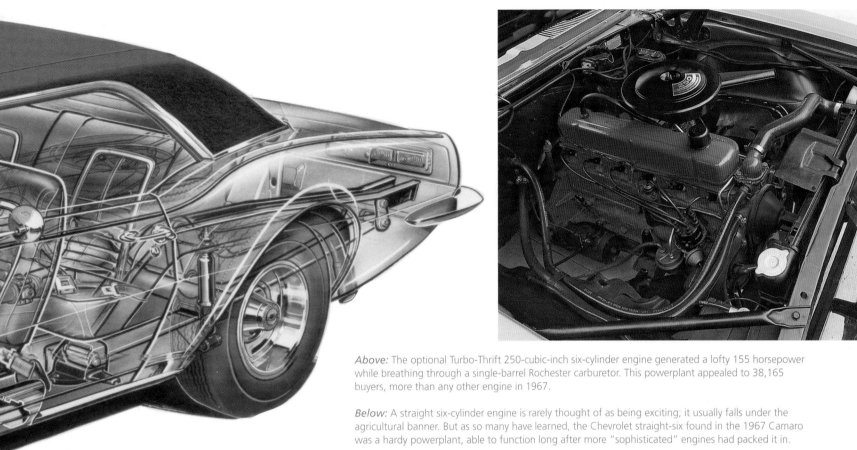

Above: The optional Turbo-Thrift 250-cubic-inch six-cylinder engine generated a lofty 155 horsepower while breathing through a single-barrel Rochester carburetor. This powerplant appealed to 38,165 buyers, more than any other engine in 1967.

Below: A straight six-cylinder engine is rarely thought of as being exciting; it usually falls under the agricultural banner. But as so many have learned, the Chevrolet straight-six found in the 1967 Camaro was a hardy powerplant, able to function long after more "sophisticated" engines had packed it in.

Camaros equipped with the optional 250-cubic-inch straight six-cylinder engine might not have set the motoring world on fire, but with the Camaro tipping the scales at only 2,910 pounds when fitted with the six-banger, the power-to-weight ratio leaned toward the sporting.

rear-axle setup was virtually stock Chevy II. A pair of single-leaf "mono-plate" springs ran longitudinally and were a bit shorter than the Chevy II's. The tubular shock absorbers were affixed outboard of the springs; base-suspension vehicles had their shocks virtually vertical, while the beefier SS machines used staggered shocks. Vehicles that packed a V-8 beneath the long hood had a radius rod attached to the passenger side of the axle in an effort to restrain wheel hop, but with the clock ticking, the engineers just ran out of time to polish the rear suspension to the same degree as the front.

One area that did get an upgrade was the brakes. The standard units were 9.5-inch drums on each corner, little better than dragging a foot out the door to induce deceleration. Buyers willing to pony up $79 could have disc brakes fitted to the front, and power-assist could be had for just $42. Metallic-lined rear brake shoes were a $37 option, and if serious braking was anticipated they could come in very handy.

Unlike the Mustang, which was offered in three body configurations, coupe, fastback, and convertible, the new F-body was going to be available in just two body styles,

A quartet of 48mm Weber downdraft carburetors sitting on a Moon induction system topped the Mark IV 396-cubic-inch big-block V-8 nestled between the Cherokee's front inner fenders. This "unstamped" pre-production L78 engine was rated at 375 horsepower before the Webers were installed. They're good for at least another 50 ponies.

coupe and ragtop. When the roof was removed to create the convertible, considerable structural rigidity was lost, and the F-body was found to have a tendency to try to shake itself apart over marginal road surfaces. The engineers tried to counteract the cyclic oscillations by a variety of means, including structural reinforcements bolted beneath the car, and beefed up door strikers. But the solution was simpler than these efforts; heavy iron weights were attached to a spring inside a steel cylinder placed at each corner. Nicknamed cocktail shakers, they dampened the torsional shake and transformed the convertible into a smooth ride. Behind each headlight was a 25-pound damper, while in the trunk, next to each taillamp, was a 50-pound shaker. Not the most elegant solution, but time was the biggest hurdle, and if it took 150 pounds of metal to enable a quick fix, then so be it.

Meanwhile, the styling of the exterior was firming up. Haga, like many in GM's styling studios, including Assistant Studio Chief John Schinella, was influenced by European designs, and the new F-body was to be the recipient of those influences. Of course, in order to fight the Mustang, the new Chevy had to have the same general proportions as the Ford, such as a long hood, short deck, and a 2+2-sized interior. These elements were incorporated into the new design, but Chevrolet's stylists took a different path than the crew from Dearborn. Where the Mustang had long, straight lines, the F-body wore taut, curving sheet metal, highly suggestive of continental offerings. The center of the car had a "pinched" waist, like a Coke bottle. This design cue was to appear on quite a number of Chevrolet products, including the 1968 Corvette, but the Camaro introduced it to Chevrolet buyers a year earlier.

On June 29, 1966, the name of the newest addition to the Chevrolet lineup was unveiled. In Larry Edsall's book *Camaro: A Legend Reborn*, the story behind the name is

Left: The 1967 Camaro Cherokee was a test bed for design and engineering work, such as the bumperettes, the molded-in rear spoiler, the transparent rear-facing hood scoop, the hood tachometer, and two-tone interior upholstery. Candy Apple Metalflake Red is laid over the cars original Aztec Gold Metallic paint, resulting in a deep and lustrous finish.

Inset: The interior of the 1967 Camaro Cherokee was primarily stock big-block Camaro. A tilt wheel sourced from a Corvette made ingress/egress easier, and the T-handle shifter fell readily to hand. Beneath it was a Turbo Hydra-Matic 400 three-speed automatic transmission, which fed power to the 12-bolt rear end. This car had heavy-duty written all over it.

Above: For buyers wanting to go whole hog on a brand-new 1967 Camaro, a convertible RS/SS was the hot ticket. With its pony car proportions, restrained styling, and wide range of performance and comfort options, the Camaro could be tailored to every purse and purpose. *Author collection*

Right: Period advertisements showed the 1967 Camaro RS/SS convertible as a lifestyle-enhancing vehicle. Evidently, if you parked your Camaro on black plastic under annoyingly bright lights, beautiful young women would cling to the car like glue. *Author collection*

revealed. "'Gemini,' 'Colt,' and 'Chevette' were considered until Chevrolet Merchandising Manager Bob Lund and Ed Rollet, a GM vice president, discovered 'Camaro' while going through decades-old foreign language dictionaries. Camaro, they read in *Heath's French-to-English Dictionary* published in 1936, meant 'friend,' 'pal,' or 'comrade.'

"'This [name] suggests the real mission of our new automobile—to be a close companion to its owner—tailored to reflect his or her individual tastes and at the same time provide exciting personal transportation,'" Chevrolet General Manager Elliot 'Pete' Estes said at the car's introduction to the automotive media on June 29, 1966. 'Chevrolet has chosen a name which is lithe, graceful, and in keeping with our other names beginning with C. It suggests the comradeship of good friends . . . as a personal car should be to its owner.'"

Not said to the media was the desire to have an almost feminine name, a non-aggressive sound that wouldn't scare off female buyers. The marketing staff had taken note of the

Ah, the good life. Driving a 1967 convertible down your long driveway gives a hint at the good times in store. Automotive advertising was undergoing considerable changes as America was entering the turbulent late 1960s. Static images like this would soon be consigned to the dustbin. *Author collection*

Right: The graceful lines of the full-width grille were maintained on 1967 RPO Z22 Rally Sport Camaros by using electrically activated doors in front of the headlights. Chevrolet priced the Rally Sport option at $105.35. A small RS badge was fitted in the center of the eggcrate grille.

Below: When the 1967 Camaro SS debuted, the 350-cubic-inch V-8 was the biggest engine in the SS model. Rated at 295 horsepower, it could sprint to 60 miles per hour in 7.8 seconds, while the engines relatively light weight helped handling. Access to all major components was easy, as there was room to spare beneath the hood.

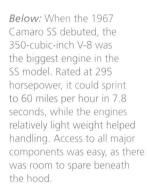

large proportion of female customers who purchased the Ford Mustang, and Chevrolet wanted some of that market.

Riding on a 108-inch wheelbase, the overall length of the F-body was 184.7 inches, only 1.5 inches longer than the Chevy II. But the layout was significantly different on the F-body than on the economy car; a short overhang in

the front was balanced by a moderate overhang at the stern. The result was a beautifully proportioned vehicle that didn't look anything like its primary rival. And the interior was just as stylish as the exterior. Following the sporty lead set by the exterior stylists, Assistant Chief Designer George Angersbach, head of Chevy #2 Interior Studio, worked to create an attractive instrument panel without breaking the bank. He wanted it to look nothing like that in a Chevy II, while appearing that it had the DNA of a Corvette. He ended up with a dual-pod design, with the speedometer in one recessed pod and a tachometer in the second pod. Standard bucket seats, the deeply inset primary gauges, and a sloping center stack of secondary controls gave the new F-body a strongly sporty look. While the rear seats were only fit for children and people you didn't like, they provided considerable utility, especially when the optional $26.35 fold-down seat was ordered.

What good is an attractive body if the car can't get out of its own way? This question was eating at the decision makers at Chevrolet, and for the most part, they followed Ford's lead in choosing what engines to offer. A six-cylinder was a must, appealing to the buyers who wanted to be seen

Above left: The tasteful rear spoiler on the 1967 Camaro SS generated some useful downforce, but more importantly, it evoked racing, a pastime that Chevrolet was participating heavily in. The company believed in the axiom "win on Sunday, sell on Monday," and any tie-in between race cars and street cars was seen as an asset.

Above right: Visible through the thin-rim steering wheel, the beefy shifter helped the driver of this 1967 Camaro SS 350 stay in the powerband. Performance cars of this era never let you forget that you were sitting in a machine, not in a rolling isolation chamber. The feedback, both aural and tactile, was part of the entire performance experience.

Left: This 1967 Camaro RS is equipped with a 327-cubic-inch V-8, and the T-handle shifter in the center console means that a two-speed Powerglide automatic transmission is filling the tunnel. This three-element torque converter gearbox used a two-speed planetary gearset and was available with any engine in the Camaro lineup with the exception of the 396-cubic-inch V-8. That engine was just too strong for the Powerglide.

in a sporty-looking car while cruising in the economy lane. With a 230-cubic-inch cast-iron straight-six that developed a breathtaking 140 horsepower, you were pretty much limited to the economy lane

Of course, options were always available—in this case, a $26.35 bump into serious power. For that money, a buyer took home a Turbo-Thrift six, the RPO (regular production option) L22 with 250 cubic inches, packing 155 ponies. A single-barrel Rochester carburetor rested on top. Available transmission choices were simple too. With the base six-cylinder engine, a three-speed manual and the two-speed

Powerglide automatic transmission were available. Customers that stepped up to the 250-cubic-inch engine could choose between a three-speed or four-speed manual box, as well as the trusty Powerglide. *Car Life* magazine tested a 250-cubic-inch six Camaro and found that it could cover the quarter-mile in a languid 18.5 seconds, tripping the lights at 75 miles per hour. But in fuel economy testing, the magazine averaged 19.2 mpg, a pretty good figure in 1967.

For the driver who wanted a bit more beans beneath the hood, there was no shortage of choices, with four V-8s

continued on page 35

Above: Chevrolet used traditional hot-rodding techniques like increased compression and a four-barrel carburetor to coax 275 horsepower out of the 327-cubic-inch L30 V-8 engine.

Above right: A large, under-stressed powerplant like the 396-cubic-inch V-8 was a durable, strong engine that could run for many thousands of miles between service. While fans of effete foreign sports cars looked down their noses at the burly mill as a simple design, its very simplicity was one of its major strengths.

Right: The Rally Sport option could be installed on any Camaro, regardless of engine, and while it didn't affect the suspension or powertrain, it did set the car apart visually from its competition, the Mustang. Chevrolet marketed the Rally Sport option as "a more glamorous version" of the Camaro. It could be fitted to both coupe and convertible models.

The handsome Rally Sport option, RPO Z22, cost dealers $76 and retailed for $104.35. The package consisted of electrically operated headlight doors, headlight washers, back up lights, front parking lights mounted in the lower valance, lower rocker panels, black taillight housings, wheel arch bright molding, and more.

One of the design features on the 1967 Camaro was the pinched waist, often called a Coke-bottle look. It was intended to reduce the visual weight of the side sheet metal and give the Camaro a sporty demeanor. It was very successful, affording the car a youthful air, perfect for competing with the Ford Mustang.

Top: Every 1967 Super Sport Camaro was fitted with the highly regarded F41 Suspension Package, which included heavy-duty shock absorbers and springs. A traction bar was fitted to the passenger side of the rear suspension in an effort to tame axle hop under heavy acceleration. Standard tires were red-stripe D70x14 two-ply bias-ply tires.

Above left: Tom McCahill of *Mechanix Illustrated* ran a 1967 Camaro SS350 through its paces, reaching 60 miles per hour in 8 seconds and covering the quarter-mile in 15.9 seconds. He reported that the top speed was in excess of 118 miles per hour. Dan Gurney, writing for *Popular Mechanics*, reached 60 miles per hour in 7.3 seconds.

Above center: Chevrolet designers felt that putting the name on the side of the fender of the all-new Camaro in 1967 was a prudent idea to educate other motorists. The 1967 Camaro was applauded for its restrained good looks with a minimum of gaudy frippery.

Above right: Chevrolet used two different bumblebee stripes for the nose of the 1967 Camaro SS/RS, which were available in either white or black. The stripes could be installed on all Camaros as RPO D91, a $14.75 option. The headlight doors on 1967 Rally Sport Camaros were electrically operated. Getting the doors to open on cue in cold weather could be a challenge.

When the 1967 Camaro was under development, the plan was to ensure the engine compartment could handle a big-block engine without any radical modifications. Chevrolet had studied what Ford had done with the Mustang and that car's inability to handle a big engine. Though the Camaro SS used a small-block when the car was first introduced, by midyear, a 396 was in the showrooms.

Above: Chevrolet had to build at least a thousand examples of the Z/28 to qualify the machine for the SCCA's Trans-Am racing series. It was full of go-fast factory parts, including a heavy-duty radiator, quick-ratio steering, 3.73:1 rear axle gears, special suspension, and dual deep-tone exhaust.

Right: A trio of engine gauges in front of the shifter might have indicated if there was a problem under the hood, but savvy drivers could depend on their hearing to pick up any hint of mechanical trouble. Fortunately, the 1967 Camaro was a simple vehicle, and it tended to be durable and reliable.

Left: The bumblebee stripe around the nose of the 1967 Camaro SS was an attempt on the part of the designers to shorten the nose visually. The electrically powered headlight doors were available only as part of the Rally Sport package, which cost $105.35 for 1967.

Above center: When the vehicle that became known as the Camaro was in development, it was known within General Motors as the Panther. Auto manufacturers often use a project name that bears no resemblance to the final name. At the time, most Chevrolet vehicles started with the letter *C*.

Above right: Unlike Camaros from a decade later, the 1967 Camaro SS 350 used very discrete badging to let the world know what the car was capable of. Instead, a heavy right foot would fling the freshman Camaro down the road, dispelling any doubts about the cars ability to run with the big boys.

on the menu. First up was the 327-cubic-inch engine, a hardy powerplant that came in two levels of tune: a base 210-horsepower version and the $92.70 RPO L30 mill that was rated at 275. Next up was RPO L48, which featured a 350-cubic-inch V-8 that was new to the Chevrolet roster of engines. Using a Rochester Quadrajet four-barrel carburetor and 10.25:1 compression, it belted out 295 horsepower. In the November 1966 issue of *Car & Driver* magazine, their crack staff wrung a quarter-mile time of 16.1 seconds at 86.5 miles per hour from an SS350.

Rear axle ratios ranged from a fuel-friendly 3.01:1 to a neck snapping 4.88:1. This is just another example of the manufacturer allowing buyers to tailor a vehicle to their specific needs and taste. It's interesting to note that the new 350 engine was available only in the Camaro for model year 1967.

If those tastes ran to more power, then the next stop on the power food chain was a race-bred engine, the 302-cubic-inch V-8 found in the Z/28. The moniker Z/28 didn't mean

anything special at the time; it was simply the next letter and number in the standard General Motors ordering system. At the time, Chevrolet was heavily involved with the SCCA's Trans-Am racing series, and the new Camaro in track configuration was limited to 305 cubic inches. In order to homologate the Z/28 for racing, Chevrolet had to build street-legal models that customers could actually take home. In order to meet the displacement regulations while holding down costs, Chevy engineers slipped the crankshaft from a 283-cubic-inch V-8 into a 327-cubic-inch block. The resulting 4-inch bore and 3-inch stroke gave the new engine 302.3 cubic inches of displacement, and with the short stroke, it gathered up revolutions faster than a politician chases votes. Chevy rated the 302 engine at 290 horsepower, which was true. It made 290 on its way to more than 400. The Z/28 vaulted to the end of a drag strip in 14.9 seconds, with 97 miles per hour showing on the speedometer.

For buyers who felt that bigger was better, the 1967 Camaro didn't disappoint. The L35 engine, displacing 396

Above: The release of the 1967 Camaro SS was also the release of the 350-cubic-inch V-8 engine. In Super Sport guise, the RPO L48 engine was rated at 295 horsepower and could propel the Camaro down the drag strip in just 15.4 seconds at 90 miles per hour. With the right gear set in the pumpkin, it could reach a top speed of 121 miles per hour.

Right: In order for Chevrolet to compete in the SCCA's Trans-Am racing series in 1967, at least 500 street versions of the potent Z/28 race car had to be built. By the time the model year wrapped up, a total of just 602 were constructed. There were no external markings on the car denoting that it was a Z/28, just a pair of hood and trunk stripes. Very sneaky.

cubic inches, was released in the Camaro in November 1966, and with a host of Corvette-sourced engine internals, this hydraulic-lifter engine was rated at 325 horsepower. This Turbo-Thrust engine stickered for $263.30, and while it weighed 186 pounds more than the 350-cubic-inch V-8, it had 410 lb-ft of torque to help vaporize the rear tires. *Motor Trend* magazine flogged one down the drag strip, coming up with a time of 15.4 seconds at 92 miles per hour. It could reach 60 miles per hour in 6.8 seconds, impressive numbers considering the tires of the day. It was a popular engine, with 4,003 sold.

Later in the model year, the Camaro was the recipient of an even hotter 396 engine. Denoted RPO L78, it was a solid-lifter beast, cranking out 375 horsepower and 415 lb-ft of twist. With a price of $500.30, an 11.0:1 compression ratio, and a huge Holley four-barrel carburetor, it was a thirsty, loud, and fast. Despite its late entry into the Camaro, 1,138 were sold. Incredibly, 396-equipped Camaros used the same

Left: The 1967 Camaro Z/28 rode on a 108-inch wheelbase. The rear suspension was made up of a live rear axle using Mono-Plate single-leaf rear springs of special uniformly stressed chrome carbon steel cushion-mounted to the axel with heavy rubber pads and attached to front and rear mounting points through rubber bushings.

Below: With its 302-cubc-inch V-8 rated at 290 horsepower, the 1967 Camaro Z/28 could haul butt down a drag strip in 14.9 seconds at 97 miles per hour, and it enjoyed a top speed of 124 miles per hour. It's no surprise that the Z/28 was fleet footed; it was the street version of a race car.

Top left: A trio of secondary gauges and a clock were mounted in front of the Hurst shifter to let the driver keep an eye on events in the engine room. Like the rest of the powertrain in a 1967 Z/28, the transmission was built strong, a valuable asset in the rough and tumble world of Trans-Am racing.

Top right: Like a freight train, the Camaro was at its best in a straight line, especially in Super Sport guise. With a big-block engine installed, the Camaro's acceleration felt like a locomotive pushing hard. But like a locomotive, stopping and rapid turns were sometimes a challenge with the heft of a big-block in the nose.

Above: This is the actual 1967 Camaro that paced the Indianapolis 500 race. Number 92, it is believed to be the first RPO L78 built. It started life as an SS396 convertible and was tagged "Special Promotional Vehicle." Constructed under Engineering Build Order No. 98168, two cars were constructed: this car for track duties, the second held in reserve.

Above: Somewhat surprisingly, the actual pace car for the 1967 Indy 500 used a Turbo Hydro-Matic three-speed automatic transmission. Granted, the tranny was disassembled and blueprinted before being installed in the car. The last thing anyone wanted was a mechanical problem with the pace car.

Left: Yellow flags mounted on the rear bumper got the attention of the race cars trailing the Indy 500 pace car in 1967. Chevrolet built a hundred replicas to be sold at dealerships to capitalize on the race exposure. Canadians were upset that they weren't included in the replica program, so Chevy built an additional 21 cars to be shipped north of the border.

Below: Chevrolet couldn't have found a better showcase for the new Camaro SS/RS than having it act as the pace car for the 1967 Indianapolis 500. The crossed flags on the front fenders were very appropriate for the job at hand.

Right: The engines in the 1967 Indy 500 pace cars were disassembled and inspected. Besides blueprinting, a camshaft and valve springs from RPO L34 were installed. After reassembly, the engines were run for 20 hours twice to bed in the reciprocating parts. The result was a *very* reliable 375 horsepower.

Below: Chevrolet depended on cubic inches, 396 of them to be exact, to keep the 1967 Camaro SS pace car in front of the field during the '67 Indy 500 race. While the engine was blueprinted, balanced, and run in, and a new camshaft was installed before the race, the powerplant was pretty much stock.

single-leaf rear suspension spring rate as the six-cylinder models. For 1967 only, both rear shock absorbers were mounted in front of the rear axle on the base cars, resulting in attention-grabbing axle hop under heavy throttle

Chevrolet didn't wait to utilize the highly regarded Super Sport (SS) moniker on the Camaro. From Day 1 (actually September 12, 1966) two SS packages were in place and ready to spice up the F-body. The first was the SS350. With a pair of simulated vents on the hood, a heavy-duty suspension, dual exhaust, special graphics, and SS badges mounted on the front fenders, grille, gas cap, and on the horn button. A "bumblebee" nose surround stripe was available, and it had a dual purpose. Besides visually shortening the length of the nose, it just looked aggressive. The second Super Sport offering came with the brutal 396 engine. The SS396 was similar to the SS350, with the exception of the huge powerplant under the long hood, as well as the deletion of "396" callouts on the badges. Borrowing the cross-flag badges from the Impala and Chevelle alerted sharp-eyed drivers that the lithe machine in the next lane was to be taken seriously.

Like all the U.S. manufacturers, Chevrolet believed that if one option is good, a lot more is a lot better. Thus the Rally Sport (RS) package was released. It didn't provide any performance enhancements; it was an appearance option.

1967

MODEL AVAILABILITY	two-door coupe or convertible
WHEELBASE	108.1 inches
LENGTH	184.6 inches
WIDTH	72.5 inches
HEIGHT	51.0 inches
WEIGHT	2,900 lbs
PRICE	$2,466
TRACK	59.0/58.9 inches (front/rear)
WHEELS	14 x 5 inches
TIRES	14 x 7.35 inches
CONSTRUCTION	unitized body/frame with bolt-on front subframe
SUSPENSION	long-arm/short-arm with coil springs front/longitudinal leaf springs, live axle rear
STEERING	recirculating ball
BRAKES	four-wheel drums, 9.5 x 2.5 inches in front, 9.5 x 2 inches in rear
ENGINE	140-horsepower, 230-cubic-inch I-6; 155-horsepower, 250-cubic-inch I-6; 210- or 275-horsepower, 327-cubic-inch V-8; 295-horsepower, 350-cubic-inch V-8; 325-horsepower, 396-cubic-inch V-8
BORE AND STROKE	3.875 x 3.25 inches (230), 3.875 x 3.53 inches (250), 4.00 x 3.25 inches (327), 4.0 x 3.48 inches (350), 4.094 x 3.76 inches (396)
COMPRESSION	8.5:1 (230, 250), 8.75:1 (210-horsepower 327), 10.0:1 (275-horsepower 327), 10.25:1 (350), 10.25:1 (396)
FUEL DELIVERY	single-barrel (230, 250), single two-barrel (210-horsepower 327), single four-barrel (275-horsepower 327), single four-barrel (350), single four-barrel (396)
TRANSMISSION	three- and four-speed manual, two-speed automatic Powerglide, three-speed automatic Turbo Hydra-Matic.
AXLE RATIO	Ranging from 2.73:1 to 4.88:1
PRODUCTION	58,808 six-cylinder, 162,098 V-8

Two actual pace cars were constructed for the 1967 Indianapolis 500, the primary vehicle and a backup. With its flags fluttering, the actual pace car glistens in the setting sun. Race winner A. J. Foyt refused the car, saying it didn't have air conditioning, and being from Texas, he needed it. As a driver for Ford, it might not have been a good idea to accept a Chevy.

Modifications to the interior of the 1967 Camaro SS that paced that year's Indianapolis 500 were limited to installing secondary gauges in front of the shifter, removing the passenger sun visor, and fitting a grab handle in its place. Most of the significant work was done on the drivetrain and suspension.

Left: The SS version of the Camaro came in two flavors: the 396-cubic-inch version, chosen by 4,003 buyers, and the 350-cubic-inch version, chosen by 29,279 buyers.

Below: Chevrolet engineer Vince Piggins was the driving force behind the Z/28, ordering a prototype built, then putting Chevrolet President Pete Estes, another engineer, behind the wheel. Estes took the car for a spirited drive and immediately gave the go-ahead to put it into production.

The stand-out feature of the Rally Sport option, Z-22, was hidden headlights. A pair of electrically actuated doors slid inboard when the lights were turned on. When closed, the doors visually widened the grille, giving it a full-width appearance. Other RS features were bright metal trim on the doorsill area, wheel openings, drip rail moldings, as well as RS badging on the gas door, grille, fenders, and steering wheel. The Super Sport option could be combined with the Rally Sport option on the same car, but the SS package would take precedence. That meant that while the car had the hidden headlights and other RS bits, all of the badging was SS. The Z/28 could also be combined with the RS package, and like an SS/RS merger, the Z/28 nomenclature was the lead option. It was not possible to combine the Z/28 and SS options.

Overall sales of the 1967 Camaro were encouraging to Chevrolet, who felt that any sales taken away from the Ford Mustang was a good thing. Buyers took 220,917 Camaros home, compared with the Mustang's total sales of 472,121 units. The Camaro delivered impressive first-year numbers against the Mustang, which had been on the market since 1964. The 1967 Camaro followed in the Mustang's tracks as well by being tapped to pace the 1967 Indianapolis 500.

1968: A TIME FOR POLISHING

With all of the effort to get the 1967 Camaro into the market to combat the Mustang, the crew at Chevrolet worked on smoothing some of the Camaro's rough edges for 1968, as well as incorporating mandated lighting and emission regulations. One of the biggest gripes was the tendency of the rear axle to wind up and release under heavy throttle, causing the axle assembly to bounce like a basketball. While this wasn't an issue on vehicles equipped with six-cylinder engines, the beefy V-8s, especially the big-blocks, had far more horsepower and torque than the chassis could handle. Slow-motion movies of the rear wheels taken by General Motors engineers at their proving grounds showed that the rear wheels moved fore and aft, as well as up and down. In an effort to at least reduce axle hop, the rear shock absorbers were staggered; the left-side shock was positioned behind the axle, while the right-side unit was in front of the axle. The single-leaf rear spring was found to be wanting in high-horsepower applications, so vehicles equipped with the L48 350-cubic-inch engine, as well as all 396-powered cars, used a multi-leaf setup. The result of all these changes was a significant reduction in undesirable axle motions. In short, it was able to get a lot more power to the ground.

continued on page 55

For 1968, the bumblebee nose stripe was modified into half a stripe, with a spear pointing toward the rear of the car, ending at the after portion of door. The body shell of the 1968 Camaro was essentially carried over from the preceding year, with minor changes to the front and rear end and overall refinement.

Above: It's good to be the boss. Chevrolet boss Pete Estes was given this 1968 convertible with all the Z/28 options for 1969, including a domed cowl-induction hood. The engineers wanted Mr. Estes to okay the changes for 1969, and as Mr. Estes was an engineer and performance enthusiast, they felt that if he was able to drive a car with the improvements, he'd be more inclined to approve them. It worked.

Right: In order to provide "sufficient" acceleration, this engineering study was equipped with aggressive 4.88:1 rear axle gears in a 12-bolt rear end; to slow the beast down, RPO JL8 brakes were fitted. This gave Mr. Estes' car four-wheel disc brakes, and to put frosting on the cake, the disc brake package was the highly regarded J56 twin-pin setup sourced from the Corvette.

Right: The experimental 1968 Z/28 convertible that was given to Chevrolet manager Pete Estes to "evaluate" performance equipment for the 1969 Z/28 was loaded with every creature comfort that the engineers could fit. Even the sound system was upgraded: The four knobs in the center console control speaker balance and reverb.

Below: Chevrolet engineers wanted their hopped-up 1968 Z/28 convertible to be as strong as possible when Chevy boss Pete Estes turned the key, and by fitting a cross-ram intake manifold onto the 302-cubic-inch V-8, and a high-lift camshaft and transistorized ignition, this potent small-block would rev like a race car.

Not for the faint of heart, the RPO L78 396-cubic-inch V-8 in the 1968 Camaro SS was a thing of brutal beauty. Rated at 375 horsepower, it was a well-engineered four-wheel missile. An open element air cleaner ensured the huge engine had plenty of fresh air to help it convert dinosaur juice into rotational energy.

Above: The original owner installed a set of traction bars to the rear suspension in an effort to control the rear wheel hop caused by the massive amount of power trying to get from the L78 396-cubic-inch engine to the narrow rear tires. It was a good effort, but the vast amount of torque tended to overcome any suspension.

Left: Matador Red, paint code RR, was a popular color on the 1968 Camaro SS 350, with two choices of interior color, black or black and white. The rarest factory-installed option in 1968 was the "Speed Regulating Device"; 0.1 percent of Camaros that year were fitted with this option.

Right: RPO C08—the vinyl roof cover—cost $73.75, and as the years worked on the 1968 Camaro SS's body, rust tended to accumulate beneath that vinyl covering. But the roof treatment gave the sporty Camaro a formal air, balancing the street-fighter looks that the red paint and SS trim imparted. The "induction stacks" on the hood were for show only and part of the Super Sport package.

Opposite top left: Three numbers on the front fender could have a chilling effect to someone pulling up to this 1968 Camaro SS at a stoplight. In a vehicle as light as the Camaro, the addition of so much power gave the Chevrolet pony car startling acceleration, the benchmark for performance in 1968. Braking and handling were given cursory attention.

Oppostie top right: A wide range of transmissions were available on the 1968 Camaro SS 350, from the base heavy-duty three-speed manual, to the four-speed manual and the two-speed Powerglide automatic transmission. The SS 350 Camaro was a good seller in 1968, with 12,496 units rolling over the curb.

Opposite bottom: The 1968 Camaro SS exemplified the term lithe, its neat lines and tidy proportions exuding a quiet purposefulness, at least until the L78 396-cubic-inch engine was fired up. Then it was time to get the hell out of the way. This car reeked testosterone.

Above: While the handsome interior was billed as a four-seater, like most pony cars, it was actually a 2+2. The rear seats were suitable for children, groceries, and people you didn't like. The large shift handle in this 1968 Camaro SS396 is connected to a three-speed Turbo Hydra-Matic automatic transmission.

Right: Deep binnacles hid the speedometer and tachometer from anyone but the driver, but having the instruments buried within the panel kept windshield reflections to a minimum at night. The horn button in the center of the steering wheel was an ideal place to affix an SS badge.

Far right: Only one automatic transmission was available for use in the 1968 Camaro SS equipped with a 396-cubic-inch V-8, and that was the beefy Turbo Hydra-Matic 400 three-speed. The "400" designation was earned from its ability to handle 400 lb-ft of torque without a problem. The crossmember used with the TH400 tranny was formed from tubing, not plate steel.

Chevrolet built 10,773 1968 Camaro SS models equipped with the 396/325-horsepower L35 option. This engine cost $263.30, a very reasonable sum for the brutish grunt the big-block generated. It would vault the Camaro down a drag strip in 14.5 seconds at 99 miles per hour in 375 horsepower guise. Top speed was 117 miles per hour.

Right: When this 1968 Camaro SS was new, the original owner immediately removed the hubcaps and tossed them into a box, which then sat in his basement. When the car was restored years later, the hubcaps were dusted off and reinstalled. In the 1960s, hubcaps were often rotating works of art, and these are a perfect example.

Below: The bucket seats in a 1968 Camaro SS didn't have much in the way of lateral support, but as this vehicle's purpose was going as fast as possible in a straight line, lateral support rarely became an issue. More important was slamming the big shifter into the next gear while keeping the gas pedal buried in the carpet. It's a matter of priorities.

Another change for 1968 was more visible than the improved rear suspension—the introduction of Astro Ventilation and the removal of side vent windows. One-piece door windows gave the Camaro's side profile a cleaner look. Other changes were apparent inside the vehicle, as the instrument panel was modified, with the primary gauges now housed in rectangular recesses. Secondary gauges mounted on the center console enjoyed the same treatment, giving the interior a more upscale look.

One change that wasn't visible but increased reliability was the actuation system used to operate the headlight doors on RS-equipped vehicles. For 1968, the switch was made from electrical motors to vacuum operation. Some owners of 1967 RSs had reported that the headlight doors could freeze shut, and the electric motors couldn't open them. The use of

continued on page 60

Above: While the suspension was based on the capable F41, Chevrolet used beefier springs and shocks to handle the considerable weight of the big-block. An SS396 was not the best car with which to tackle a sinuous canyon road, but when the blacktop was straight, it was a formidable car.

Below: If a driver pulled up to a 1968 Camaro SS and overlooked the SS badging, it would be hard to miss the simulated carburetor stacks attached to the hood. Though the bright bits weren't functional, they did indicate that the engine under the hood was nothing to ignore.

Right: Soon after the Camaro debuted in 1967, it was the lucky recipient of the race-bred 396-cubic-inch big-block V-8. This engine carried over to the 1968 model year, where it was available with three outputs: 325, 350, and 375 horsepower.

Below: Chevrolet asked Smokey Yunick to sort out a hemi-head small-block engine for possible motorsports application, and as payment, Yunick got all three development engines. In 1977, he installed one of the engines in a former 1968 Z/28 race car. It took nine years, but the race engine was finally in a race car.

Smokey Yunick's slogan was "Best Damn Garage in Town," and there was considerable truth in that. Chevrolet came to Yunick numerous times to solve engineering problems in an unconventional fashion. An inventor and ace racing mechanic, Yunick tended to think outside the box, finding solutions where nobody else could.

Above: Externally identical to a standard 1968 Z/28, Smokey Yunick's vehicle was devoid of any markings that might have indicated that its engine compartment was filled with a race engine. Astro-Ventilation made its debut in 1968, relegating the door vent window to the history books.

Opposite top: With a 396-cubic-inch V-8 under the hood, a 1968 Camaro SS could be optioned from mild to wild. Some of the options that serious hot rodders tended to spring for included the Posi-Traction rear axle, allowing both rear tires to leave long black marks. Another favorite was RPO H05, a 3.73:1 ratio gear set, costing a staggering $2.15.

Oppposite bottom left: While a vinyl roof may not seem like a performance option—and it wasn't—it was a popular stylistic addition. The $73.75 option was fitted to 77,065 Camaros in 1968. This was also the first year for government-mandated side marker lights.

Opposite bottom right: American automotive design in the 1950s and 1960s embraced the use of brightwork, and the 1968 Camaro Z/28 struck a balance between too much and not enough chrome. The tasteful bits accenting the wheel arch and the sidelight surround were masterful exercises in restraint. Even the Z/28 badge was made of metal, not a cheesy decal.

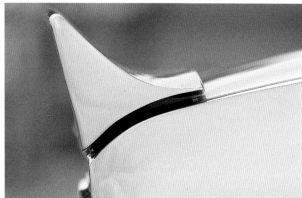

Above: The 396-cubic-inch V-8 mounted in the 1968 Camaro SS wasn't called a big-block for the fun of it; it was a big engine. The huge chromed valve covers hid a robust valvetrain that took up a lot of room. Everything on this engine was large, including its thirst.

Right: Rear spoiler RPO D80 cost $32.65 and could be fitted to any Camaro. The spoiler generated actual downforce, as well giving this 1968 Camaro SS396 an aggressive tone. The spoiler was attached to the trunk lid, and care had to be taken when shutting the trunk to avoid pinching fingers between the ends of the spoiler and the fender.

Opposite: In 1968, the panel between the taillights of the Camaro SS models were painted black, regardless of the vehicle color. Standard Camaros used a body color panel.

a vacuum system "borrowed" from the Corvette solved the problem. The grille itself was changed for 1968; instead of the flat face found on the '67, the Camaro's sophomore year had a slight V shape designed into the grille. This stylistic motif was to be more pronounced the following year.

In the name of safety, the government had decreed that starting in model year 1968, all passenger cars sold in the United States would be equipped with front- and rear-side marker lights. The Camaro used rectangular lenses—amber in the front, red in the rear. Changes were also evident on the hood, provided you ordered a big-block engine. If the big-block engine, nicknamed the Rat, was in the car, simulated carb stacks were mounted on the hood.

Beneath the hood, it was pretty much the same lineup as 1967. One significant change was the availability of two versions of the 396 powerplant. Both were rated at 375 horsepower, with 415 lb-ft of torque. Both had 11.0:1 compression, both used solid lifters, and neither was

available with an automatic transmission. The difference lay in the cylinder heads. RPO L78 used cast-iron material, while RPO L89 came with aluminum heads. They were an expensive option ($868.95) and rare too. Only 272 sets were sold. This option was an effort from the factory to reduce weight—a tacit acknowledgment that the primary purpose of the big-block was to cover as much ground as quickly as possible while traveling in a straight line. In some circles this is known as drag racing.

Sales for the 1968 model year surpassed the Camaro's freshmen outing, with 235,147 units sold. The coupe model was the more popular, selling almost 10:1 against the convertible. It was clear that the public was still smitten by pony cars in general, and the Camaro in particular. Part of the Camaro's appeal was the ability to tailor it from the dealer, creating a car that could be mild or wild, or anywhere in between. Yet the most successful selling version of the first-generation Camaro started rolling down the assembly lines on September 26, 1968, and would, over time, become the seminal Camaro.

Built to grace the show circuit, the 1968 Camaro Caribe Concept was an open-air two-seat version with a huge open area where the rear seats and trunk would normally be. This truck-like bed was built to gauge public reaction, which was probably rather tepid. Note the custom windshield frame.

1968

MODEL AVAILABILITY	two-door coupe or convertible
WHEELBASE	108.1 inches
LENGTH	184.7 inches
WIDTH	72.6 inches
HEIGHT	51.5 inches
WEIGHT	2,950 lbs
PRICE	$2,565
TRACK	59.0/58.9 inches (front/rear)
WHEELS	14 x 5 inches
TIRES	14 x 7.35 inches
CONSTRUCTION	unitized body/frame with bolt-on front subframe
SUSPENSION	long-arm/short-arm with coil springs front/longitudinal leaf springs, live axle rear
STEERING	recirculating ball
BRAKES	four-wheel drums, 9.5 x 2.5 inches in front, 9.5 x 2 inches in rear
ENGINE	140-horsepower, 230-cubic-inch I-6; 155-horsepower, 250-cubic-inch I-6; 210- or 275-horsepower, 327-cubic-inch V-8; 295-horsepower, 350-cubic-inch V-8; 325–375-horsepower, 396-cubic-inch V-8
BORE AND STROKE	3.875 x 3.25 inches (230), 3.875 x 3.53 inches (250), 4.00 x 3.25 inches (327), 4.0 x 3.48 inches (350), 4.094 x 3.76 inches (396)
COMPRESSION	8.5:1 (230, 250), 8.75:1 (210-horsepower 327), 10.0:1 (275-horsepower 327), 10.25:1 (350), 10.25:1 (396)
FUEL DELIVERY	single-barrel (230, 250), single two-barrel (210-horsepower 327), single four-barrel (275-horsepower 327), single four-barrel(350), single four-barrel (396)
TRANSMISSION	three- and four-speed manual, two-speed automatic Powerglide, three-speed automatic Turbo Hydra-Matic.
AXLE RATIO	Ranging from 2.73:1 to 4.88:1
PRODUCTION	50,969 six-cylinder, 184,178 V-8

Above: The 1968 Camaro Cherokee was built to pace races, and it's seen here leading a starting grid at the Can-Am race at Road America in 1968. With five adults in the car, having a healthy 396-cubic-inch Mark IV V-8 helped keep the customized Camaro in front of the field. Famous race car driver Stirling Moss is at the wheel.

Left: Every 396-cubic-inch V-8 that was installed in the 1968 Camaro SS wore this sticker denoting that it had been built at the famed Chevrolet engine assembly plant at Tonawanda, New York. This was a source of considerable pride for car owners and a slap to the competition.

The 1969 Camaro SS is arguably the iconic model of the marque. Surprisingly, it was a single-year design, yet it sold very well. Total Camaro sales for '69 were 243,085 units, a stunning number for a vehicle that had been on the market less than three years. Clearly, people respond to superb design.

1969

Going into its third year, the Camaro's success led Chevrolet to make some subtle changes to the body. Fresh sheet metal kept the Camaro competitive with the Mustang, which wore re-styled bodywork as well. One camp in Chevrolet Styling wanted to create a lower, more aggressive stance, but word came down that the 1969 Camaro would have to retain the Nova platform, with its high cowl and relatively upright windshield. An all-new Camaro was in the works for 1970, and Chevrolet was loath to invest considerable money in a single-year vehicle that would have a production run of 18 months, maximum. So the stylists were told to make do on the existing architecture, creating a design that touched on the previous years, but would

Above: Sales brochures were, by 1969, using a more stylistic approach to showing automobiles, especially performance models. While many images of the day showed the car in an ideal setting, strong photography highlighting the vehicle's brute strength helped bring enthusiasts into showrooms. *Author collection*

Left: How thoughtful; the back of the 1969 Camaro sales brochure included a checklist that a potential buyer could mark off, then take into a Chevrolet dealership and hand the list to a salesman, hastening the purchase. You didn't want to waste any time getting behind the wheel! *Author collection*

Above: Large two-page, full-color pages in the 1969 Camaro sales brochure helped potential buyers visualize what "their" Camaro would look like. While base model Camaros could be purchased, it was a rare dealer who actually had one in stock. The dealers made a significant amount of money on options. *Author collection*

Above: For a real-world, drive-every-day muscle car, you'd be hard-pressed to beat a 1969 Camaro SS/RS 350. With a base 350-cubic-inch V-8 rated at 300 horsepower, it wouldn't have any problem merging onto a freeway, and the hidden headlights of the Rally Sport option gave the front end a tidy appearance. Factor in a convertible sunburn, and it's an unbeatable combination.

Right: The chin spoiler was a rudimentary attempt at controlling aerodynamics and was the beginning of a relationship between designers and aerodynamicists. The days when a styling team had absolute say in the final look of an automobile were drawing to a close. Yet in 1969, the front bumper was still virtually worthless at fending off minor collisions.

introduce a number of visual elements that would take the Camaro into the future. The resulting car was a grand slam.

A new front end enjoyed a deeply recessed grille, which had a more pronounced V than in previous years. The full-width bumper curled up at the ends to meet the circular line formed by the grille surround, creating a loop effect. This dramatic look was well served by both exposed headlights and the hidden lamps of the Rally Sport option. At the leading edge of the front fenders were engine call-outs denoting the engines displacement, with the exception of the base engine. A variety of graphic packages were available, including the wrap around "bumblebee" stripe that enveloped the nose.

One of the more striking additions to the 1969 Camaro was the handsome cowl-induction hood. Years of racing had shown Chevrolet that the high-pressure area at the base of the windshield was an ideal place to feed cool air into the air cleaner. This $79 optional hood (RPO ZL2) was available for installation on all SS and Z/28 models. A fiberglass version was offered over the counter for owners using a dual-quad cross-ram or single four-barrel induction system.

Above: This is how a large number of 1969 Camaros were optioned; vinyl roof, whitewall tires, and hubcaps. Contrary to what some say, not every Camaro in 1969 was a ZL-1 or a Z/28. The vast majority of Camaros were built for commuter use, but with a sporty flavor. *Author collection*

Above: In 1969, Camaro buyers who had a penchant for performance demanded to know as much about the vehicle's mechanical specifications as possible. Chevrolet responded by stuffing the sales brochure with a wealth of technical data, giving the go-fast crowd plenty to drool over. *Author collection*

Right: While they were eye candy, options were a major source of income for both Chevrolet and its dealers. The markup on options was enormous. Many options were installed at the factory, while others would be fitted at the dealership just before handing the vehicle off to its new owner. *Author collection*

SS: ALL-MUSCLE, ALL-CAR, WITHOUT A TACKY PIECE OF GINGERBREAD ANYWHERE.

Nobody else can come close to Camaro SS. Fat tires. Power disc brakes. Unique hood. Sport striping. Special suspension. Big power. Special 3-Speed transmission with floor shifter. Camaro SS: what the younger generation is coming to because it's about as close as you can come to a pure sports car and still get a back seat and trunk in the deal. We have only one more thing to say: drive it!

Left: By 1969, Chevrolet made no attempt to sugarcoat the fact that the Camaro could be optioned into a brutal road warrior. Unconventional camera angles were used in an attempt to appeal to a younger demographic of buyers. *Author collection*

Below: A painted front bumper was an option in 1969, and this earth-tone-colored 1969 Camaro was shown in an appropriately earthy setting. How the tires didn't get dirty is still a mystery. *Author collection*

WHO NEEDS TO SAY "NEW" OR "BETTER"?

Camaro . . . Chevrolet's incomparable Hugger with the wide-stance grip. Longer, wider, tougher, even quieter for 1969. Decide on a Convertible or Sport Coupe and personalize it from a long list of options and packages. A starter: new Color-Matched resilient bumper that looks like it's part of the car, yet shrugs off nicks and bumps. But put it all together the way *you* want it.

Below: The public embraced the 1969 Camaro in larger numbers than in previous years, as 243,085 units were sold for this model year. It was the first year that the Camaro offered variable-ratio power steering. The standard engine in the 1969 Camaro SS was the 300-horsepower, 350-cubic-inch V-8.

Above left: A set of "vents" embossed into the sheet metal of a 1969 Camaro in front of the rear wheel opening was intended to resemble brake cooling air vents. For cost considerations, a steel wheel was standard, but when teamed with a stylish center cap and trim ring, the rolling stock was rather handsome.

Above right: Faux "velocity stacks" graced the hood of a 1969 Camaro SS350, suggesting a race-inspired engine. A center "spine" running down the hood had a dual purpose, visually breaking up the large expanse of flat metal, as well as stiffening the hood.

Left: Unlike Ford's trio of first-generation Mustang body designs, the Camaro was offered in two body styles: coupe and convertible. Both offered brawny good looks that have kept the Camaro one of the most desired vehicles ever built. The 1969 SS Camaro incorporated faux carburetor inlets on the hood, while the COPO Camaro used a ducted hood.

RALLY SPORT: THIS IS THE
SPIRITED WAY TO
CHALLENGE A ROAD. JUST
ASK THE KID WHO OWNS ONE.

One look and you can see what all the
hullaballoo's about. Concealed headlights
and a special grille. Rear fender louvers.
Headlight washers (that's right, *headlight*
washers). You can see why we added a
maximum security locking system on the
steering column that makes it awfully
tough to steal your car. Think we outdid our
competition with this one? Good thinking.

Above: Chevrolet was thrilled to be able to put Camaro in front of the race cars at the Indianapolis 500 twice in three years. For its sophomore appearance, the pace car packed a burly big-block engine and F41 suspension to keep it in front of the snarling race cars. *Author collection*

Right: The doors in front of the headlights had slots in them to allow some light to shine in front of the 1969 Camaro RS in case the doors, for whatever reason, didn't open when the lights were turned on. The circular lights beneath the bumper were the turn signal indicators. *Author collection*

In profile, the 1969 Camaro was more flowing than its predecessors, with character lines streaming rearward from the wheel openings. Faux vents in front of the rear wheel arches hinted at race-ready sportiness. The rear treatment was tasteful, with the taillights increasing in width, and the entire taillight panel being slightly recessed. From any angle, the 1969 Camaro exuded more testosterone than in previous years. And the public loved it. With total sales of 243,085 units, it was the most successful Camaro to date. The two assembly plants that built the Camaro, in Van Nuys, California, and Norwalk, Ohio, were running multiple shifts, and for the second time in three years, the Camaro found itself leading the race cars at the Indianapolis 500 around the track.

As the Camaro got a little older, Chevrolet knew that it had a good thing going. Like the Mustang, the Camaro had

continued on page 76

Above: Dover White with Hugger Orange stripes, the 1969 Camaro pace car was popular, with 3,674 sold. Engine choices ranged from 300 to 375 horsepower. Once a buyer checked RPO Z11, "Indy 500 Pace Car Accents," mandatory options entered the picture, including D80 spoiler equipment, Z22 Rally Sport package, and ZL2 air induction hood.

Left: The base engine for the 1969 Camaro Indy 500 pace car was the SS350 package, with a three-speed manual transmission. It was up to the customer to outfit the vehicle to individual taste, including upgrading to a bigger engine. Approximately 20 percent of pace cars were equipped with the 396-cubic-inch engine; the actual Indy 500 pace car and its backup used an L89 aluminum-head 396.

It's said that the mark of a good designer is knowing when to lift the pen, and the 1969 Camaro is proof of that statement's inherent truth. A convertible like this is made for big skies and long roads. The rear spoiler was developed for Trans-Am racing duty, and it looked perfect on the Camaro SS.

Left: In the hunt for more speed, Chevrolet offered RPO L89 for the 1969 Camaro SS, a 396-cubic-inch engine topped with a pair of aluminum heads. The horsepower rating of 375 didn't change from the iron-head version, but it was felt that the weight savings were worth the $710.95. Only 311 were sold.

Left: The trunk of the 1969 Camaro wasn't the sort of space that could handle all the luggage on a family summer vacation, but for a couple's weekend trip, the Camaro was ideal. Add a folding top, and the journey promised to be memorable.

Above: Buyers willing to shell out nearly $1,000 over the Camaro's purchase price could receive a version of the solid-lifter 396 engine with aluminum L89 heads. These L89 engines generated the same 375 horsepower and 415 lb-ft of torque as the L78 engines, but the aluminum shaved off pounds.

Above: Houndstooth cloth upholstery gave any 1969 Camaro an upscale look. This COPO 9561 example was fitted with RPO Z23, which resulted in the steering wheel wearing a wood-grain accent, an assist grip above the glove box, and bright metal trim on the pedals.

Right: A tire dealer's dream. With healthy power beneath the hood, sporty looks, a raucous exhaust note, and narrow tires, the Camaro almost begged to indulge in behavior frowned upon by drivers of more mainstream vehicles, to say nothing of the reaction from law enforcement officials; they often say the Camaro as an easy way to fill a monthly quota.

Opposite: The top of the air cleaner assembly on this 1969 COPO 9561 Camaro was ringed by a rubber gasket, which mated with the bottom of the hood. Within the hood was a rear-facing scoop that allowed the high-pressure air that built up at the base of the windshield to flow into the induction system.

Right: The 1969 Camaro COPO 9561 carried a 427-cubic-inch V-8 behind the deeply inset, full-width grille, a styling element that would be brought into the twenty-first century on the 2010 Camaro. Good design is timeless.

Below right: Handsome 14x7 wheels were part of RPO Z27, the Super Sport package. This option was value-rich, costing only $295.95, and included such goodies as power front disc brakes, hood insulation, special hood, suspension and trim, and more.

an option list as long as your arm. The engine roster carried over from 1968, with a couple of additions. The base engine on Super Sports was still the RPO L48 350-cubic-inch V-8, now rated at 300 horsepower. A 225-horse version of that engine, LM1, was available in base Camaros for $52.70. Next up the 350 ladder was RPO L65, costing $21.10, and rated at 250 ponies. The remainder of the engine lineup was the same as 1968.

Later in the model year, the Camaro was the recipient of a special powerplant: the ZL1. Displacing 427 cubic inches, it was built with an aluminum block, heads, and intake manifold. It was originally intended for use in Can-Am race cars but, due to its cost, Chevrolet was amenable to letting some engines filter into the mainstream as a way of recouping some engineering expenses. As it was, there was a little-known path to higher performance called the Central Office

Above: It's clear that the original buyer valued performance over appearance. The use of "poverty," or "dog dish," hubcaps showed that money was spent in other areas, such as under the hood. This 1969 Camaro SS/RS was equipped with the rare L89 option. But note that it also sported a vinyl roof.

Left: This is the view most people saw of an aggressively driven 1969 Camaro COPO 9561. Its huge 427-cubic-inch engine would hurl the F-body toward the horizon at an alarming rate. Note the lack of identification regarding what was under the hood; this was the perfect sleeper.

Above: The 1969 Camaro Z/28 was ostensibly rated at 290 horsepower. Sure, it put out 290 horsepower—on its way to almost 400. The rules regulating Trans-Am race cars for the 1969 season limited cubic-inch displacement to 305, and the Z/28 displaced 302. But it would rev to dizzying speeds.

Right: Back in the days when American engines were measured in cubic inches, it was a point of pride to have the displacement displayed. On the 1969 Z/28, the raised portion of the cowl-induction hood served as an ideal place to mount the call-outs. Unknowing drivers would see these and think the Z/28 was underpowered. They only thought that once.

Production Order (COPO). If a dealer knew the right codes, the factory would build a vehicle equipped with options outside the norm. The ZL1 engine certainly qualified as being outside the norm. Ostensibly rated at 430 horsepower in an attempt to reduce the chances of the insurance industry flat-out killing the engine option, the ZL1 actually made better than 500 horsepower. Far better. On a dyno with open headers, the readout said 575. It was a pure race engine than was intended to add mileage to the odometer in quarter-mile increments. With a tank full of racing fuel, it could lunge the length of a drag strip in 10.41 seconds at 128 miles per hour. Not bad for a factory street legal car.

The 1969 Z/28 looks great from any angle, and from above, the amount of inset of the front grille is evident. A yellow houndstooth cloth interior was rare, the vast amount of Z/28s being fitted with a black-and-white houndstooth treatment. The slender front bumper was a graceful styling element, but did little to protect the front end.

Right: The 1969 Camaro SS L89 was the perfect blend of visual and performance punch. A surprising number of buyers—100,602—spent the $84.30 for a vinyl roof. Squaring the shape of the wheel openings for the 1969 model year gave the Camaro a more aggressive look.

Above: You had to be serious about performance to pony up $710.95 for RPO L89, which meant aluminum heads atop a 396-cubic-inch V-8. The expensive option didn't add a single horsepower to the engine, but it did remove weight. On a drag strip, that's as valuable as finding power.

Above right: Buyers of the 1969 Camaro Z/28 had a strongly styled vehicle to begin with, but for the customer who couldn't say no to tasteful options, RPO Z22 beckoned, the Rally Sport Package. The most visible part of the package was the hidden headlights. A pair of vacuum-actuated doors covered the headlights when they were off, but slid open to expose them when needed.

Right: No performance car worth its high-test came without a Hurst shifter, and this 1969 Z/28 RS was no exception. Strong, durable, and the best on the market, its positive engagement left no doubt about being in gear. Less useful were the quartet of secondary gauges on the center console. They were challenging to read in a hurry, though they looked great.

Fred Gibb, a Chevrolet dealer in La Harpe, Illinois, contacted the COPO office about installing the fire-breathing engine in the Camaro. Gibb had experience dealing with the COPO system when he ordered 1968 Chevy II Novas equipped with 396/375 engines. When he told the COPO office that he wanted the new aluminum 427 (COPO 9560AA) in the F-body, Chevrolet told him that he'd have to order at least 50 cars. Gibb placed the order for 50, and soon transporters started delivering the cars. But when Gibb saw the invoice, he choked. He was being charged $7,269 per car! When he placed the order, he had been told that each car would cost $4,900. Between Gibbs placing his order and receiving the cars, General Motors had decreed that

Above: Besides the hidden headlight doors, the Rally Sport Package included special striping, headlight washers, simulated rear fender louvers in front of the rear tires, roof drip moldings on coupes, RS emblems placed throughout the vehicle, and back up lights below the rear bumpers.

Below: With its back up lights below the bumpers, brightwork surrounding the taillights, and chrome trim highlighting the faux louvers in front of the rear wheelwells, this 1969 Z/28 was a lucky recipient of the Rally Sport Package. Costing $131.65, the option delivered a lot of content for the money.

In 1969, you could still see the engine in an automobile. Because of this, manufacturers made the effort to dress up their high-performance offerings. The 1969 Z/28 wore finned aluminum valve covers, an attractive enhancement to the looks of the engine compartment as well as allowing some cooling of the lubricating oil.

This trio of Camaros terrorized Trans-Am races in the late 1960s, and they were up to their old tricks at the 2010 Monterey Historic Automobile Races at Mazda Raceway Laguna Seca. Former racer Roger Penske enjoys a well-deserved reputation for his attention to detail, and these race cars were no exception. They were always the class of the field and frequent winners.

development costs had to be passed onto customers, and the price of a ZL1 skyrocketed. Gibb asked Chevrolet for help and, in an unprecedented move, they bought back the bulk of the cars.

The top salesman at Gibb Chevrolet, Herb Fox, was a drag racing enthusiast and had gotten Fred Gibb interested in competing. He remembers the situation. "We got thirteen sold, but we didn't know what we were going to do with the rest of them," said Fox. "We pulled into Detroit for the drags, and Gibb got called up to the tower. When he came down from the tower, I asked him what happened. He said that the general manager there thought it [the ZL1] was the

As an all-purpose muscle/road/strip machine, it was hard to top the 1969 Z/28. With its intoxicating blend of power, handling prowess, superior brakes, and devilish good looks, it was one of the finest road cars to roll out of Detroit. Born on a racetrack, the Z/28 was a popular model, with 20,302 customers snapping up the 290-horsepower thrill ride.

hot deal and wanted one. I told him to come to La Harpe and take them. He came with a semi and took all but one. We had it two years before we ever got rid of it." Only 69 ZL1 Camaros were built, putting the uber-drag car in the category of one of the rarest models made. Each one was built at the Norwood, Ohio, plant. And wouldn't you love to have been the employee that drove them off of the assembly line?

Another high-performance Camaro turned heads, usually on racecourses with turns. The SCCA's Trans-Am series was one of the most hotly contested contests in motorsports, with *heavy* manufacturer participation. Detroit's Big Three waged war on the tracks to win the battle in the showroom. Chevrolet was in the thick of the battles with its able Camaro Z/28. Regulations required that carmakers entered in the race series make street-legal cars that Joe Q. Public could buy. So all it took was desire and a checkbook to put a muscle car into the garage that could handle the curves like few other vehicles of its day. This was due to a superb mix of parts and know-how.

Kendall
GT-1
High
Performance
Motor Oil

BLAIR'S
SPEED SHOP
PASADENA, CA.

2771

MARINE ENGINE BUILDING

Above: Performance enthusiasts tended to want to work on their machines to "enhance" them, and that usually meant developing a relationship with the counterman at the local speed shop. Establishments such as this could supply the horsepower-hungry with a vast selection of parts designed to burn fuel at an ever-faster rate.

Left: Compared with some of its contemporaries, the 1969 Camaro Z/28 didn't have the massive cubic inch displacement that resulted in impressive quarter-mile numbers. Instead, the Z/28 used its high-revving 302-cubic-inch V-8 to generate sufficient power to threaten big-block cars and its race-honed suspension to outdrive them.

Right: Chevrolet mounted a set of secondary gauges in the 1969 Z/28 at the forward end of the center console so a driver could keep tabs on the high-performance engine. The gauges might not have been the easiest to read, but they did let a driver know if the engine was going south.

Above: Chevrolet, working with Smokey Yunick, developed an aluminum twin-carburetor intake manifold that unleashed another 25 horsepower above the factory-rated 290. Using a pair of 600-cfm Holley four-barrel carburetors, each carb would feed fuel to the cylinder bank opposite it. Chevrolet dealers installed the package, which cost $500.

Right: At first glance, the air cleaner sitting on top of the 302-cubic-inch V-8 in this 1969 Camaro Z/28 might seem a bit distorted, but it rested atop a pair of Holley four-barrel carburetors. The idea was that long intake manifold runners would help generate more power in the upper range of engine operation. True enough, but low-speed tractability suffered.

Left: The houndstooth cloth upholstery was part of RPO Z87, the Custom Interior package. Other upgrades included with the option included genuine simulated wood grain on the steering wheel and dashboard, bright accents on the pedals, molded door panels, a glove box light, and a trunk mat.

Below: Houndstooth upholstery added a bit of visual style to the basically spartan interior of the 1969 Z/28. Like all good sports cars, the purpose of the vehicle was to perform well in all three arenas—acceleration, braking, and handling—and the Z/28 was highly competent in each area.

1969

MODEL AVAILABILITY	two-door coupe or convertible
WHEELBASE	108.1 inches
LENGTH	186 inches
WIDTH	74 inches
HEIGHT	51.6 inches
WEIGHT	3,005 lbs
PRICE	$2,621
TRACK	59.6/59.5 inches (front/rear)
WHEELS	14 x 7 inches
TIRES	E78-14
CONSTRUCTION	unitized body/frame with bolt-on front subframe
SUSPENSION	long-arm/short-arm with coil springs front/longitudinal leaf springs, live axle rear
STEERING	recirculating ball
BRAKES	four-wheel drums, 9.5 x 2.5 inches in front, 9.5 x 2 inches in rear, front Disc and four-wheel disc optional
ENGINE	140-horsepower, 230-cubic-inch I-6; 155-horsepower, 250-cubic-inch I-6; 210-horsepower, 327-cubic-inch V-8; 290-horsepower, 302-cubic-inch V-8; 200-horsepower, 307-cubic-inch V-8; 255-horsepower, 350-cubic-inch V-8; 325- or 350- or 375-horsepower, 396-cubic-inch V-8; 450-horsepower, 427-cubic-inch V-8 (ZL1)
BORE AND STROKE	3.875 x 3.25 inches (230), 3.875 x 3.53 inches (250), 4.00 x 3.25 inches (327), 4.00 x 3.00 inches (302), 3.875 x 3.25 inches (307), 4.0 x 3.48 inches (350), 4.094 x 3.76 inches (396), 4.312 x 3.65 inches (427)
COMPRESSION	8.5:1 (230, 250), 11.0:1 (302), 9.0:1 (327, 255-horsepower 350), 10.25:1 (300-horsepower 350, 396), 12.0:1 (427 ZL1)
FUEL DELIVERY	single-barrel (230, 250), single two-barrel (327), single four-barrel (350, 396, 427)
TRANSMISSION	three- and four-speed manual, two-speed automatic Powerglide, three-speed automatic Turbo Hydra-Matic
AXLE RATIO	Ranging from 2.73:1 to 4.10:1
PRODUCTION	36,248 six-cylinder, 206,837 V-8

Buyers of the Z/28 option got a lot of technology for their money, including special front and rear suspension, a heavy-duty radiator and temperature-controlled fan, dual exhaust, quick-ratio steering, 302-cubic-inch V-8, and a bevy of Z/28 badges. A manual four-speed transmission and front disc brakes were mandatory, the option was only available on the coupe, and Posi-Traction rear axle was strongly recommended. The engine was a work of mechanical art, filled with forged race-grade parts and able to wind to 7000 rpm without the internal parts becoming external. The factory rated the engine at 290 horsepower, but with a heavy right foot it would crank out 375. A huge 800-cfm Holley double pumper carburetor fed the aluminum high-rise intake manifold, where the large-valve heads got the job done. For buyers wanting even more, a cross-ram intake manifold could be purchased over the counter. It worked with a pair of Holley 600-cfm double pumper carbs. The result was racetrack-grade power. The RPO Z28 option cost $458.15 when it was introduced in September 1968, but by November 1969, a couple of months before the end of production, the option price had risen to $522.40. This was due to increased content in the option, which was passed on to consumers.

The result of all this was a street car that could embarrass the vast majority of muscle cars in a straight line, and demolish the rest in the turns. At the drag strip, a 1969 Z/28 would cover the 1,420 feet in the mid-15-second

Left: The 1969 Z28 was a $458.15 option that included a 302-cubic-inch V-8 engine, dual exhaust with deep-tone mufflers, special front and rear suspension, rear bumper guards, heavy-duty radiator and temperature-controlled fan, quick-ratio steering, 15x7 rally wheels, E70x15 tires, 3.73:1 ratio axle, and rally stripes.

Below: Regardless of the number of carburetors sitting on top of the 302-cubic-inch engine in the 1969 Camaro Z/28, a cowl induction system fed cool ambient air into the induction system via a rear-facing hood scoop. It directed air from the high-pressure zone at the base of the windshield into the air cleaner.

range, crossing the line in the mid-90s. With its race-tested suspension, it could hustle like a sports car. Ford saw what a capable vehicle Chevrolet had developed, and its response met the Z/28 both on and off the track. The Boss 302 Mustang was a highly capable car, and the Ford and the Chevrolet were evenly matched. On a racetrack, the driver was usually the determining factor in deciding who won. The buying public loved the spectacle that was Trans-Am racing, and the excitement on the track found its way into Chevy showrooms. Sales of the Z/28 for 1969 were almost triple from the year before, settling at 20,302. Granted, with the

introduction of an all-new Camaro for 1970, Chevrolet kept the 1969 production line open longer than would be normal. But the Z/28, a specialty vehicle, sold in far greater numbers than Chevrolet had anticipated.

With the "new" Camaro on the horizon, Chevrolet extended the model year build span to 17 months. This meant that 1969 Camaros were still for sale in the early months of 1970. That inflated the sales totals for the model year, but you can't sneeze at 243,085 units. The Camaro had become more than just Chevrolet's answer to the Ford Mustang; it was an icon in its own right.

For the buyer of a 1969 Z/28 wanting even more power, a dealer-installed option was available, the Cross-Ram Induction System. Costing $500, it consisted of an aluminum intake manifold topped with a pair of 600-cfm Holley four-barrel carburetors. This package added 25 horsepower to the potent 302-cubic-inch engine's output.

Above: Released to legalize Chevrolet's involvement in the SCCA Trans-Am racing series, the Z/28 was a serious road racing machine that you could legally drive on the street. The combination of a lightweight, high-revving V-8 in a Camaro with a race-bred suspension resulted in a legendry car in its own day.

Right: Not all 1969 Z/28s were painted in vivid, attention-grabbing hues; some were more sedately finished, but no less handsome. The twin stripes played a functional role on race cars by reducing reflections into a race car driver's eyes. Chevrolet carried them onto its sportier street cars to denote race-influenced products.

Right: As in prior years, Chevrolet painted the rear panel between the taillights black on the Camaro SS396. The gas cap was located behind the rear license plate; the plate was hinged on the bottom, and a small spring would keep it vertical. Pump jockeys would use the cap to hold the plate down while pumping the high-test.

Below: The L89 aluminum-head option was pricey at $710.95, but such a price guaranteed that you'd be the only one on the block with such an exotic powerplant. Only 311 L89s were ordered, and when the engine was cold, it popped and snapped until the heads had heated and sealed.

GENERATION TWO

1970-1981 European Flair, American Power

Confidence. Chevrolet had plenty of it when the 1967 Camaro hit show-rooms. How else could you explain Styling and Engineering staffs diving into the task of creating the second-generation Camaro? Designing a vehicle took years in the days before massive supercomputers. So it was not unusual for the teams responsible for making a vehicle ready for production to toil away for three or four years before the final product rolled off the assembly line. Thus, at the same time that the public got their first look at the 1967 Camaro, the crew at Chevrolet picked up their slide rules and #2 pencils and set about designing the new car's replacement.

Within Chevrolet it was understood that the first-generation Camaro was a quick-response effort to combat the Mustang. That meant that Chevy grabbed as much technology off the shelf as it could in an effort to get the car on the road as quickly as possible. Using the Chevy II platform saved considerable time and money, but a sporty car it wasn't. And the Camaro had become a very sporty car, posed next to the Corvette in many advertisements.

Over-the-shoulder visibility was compromised by the large B-pillar that the steeply raked rear window required, but few debated that it was a great design. The new version of the Camaro was 2 inches longer and 1.1 inches lower than the 1969 model.

93

While the second-generation Camaro retained the front subframe design of the original, the frame spars extended farther back under the unit-body to provide a stiffer structure. *General Motors 2012*

Bill Mitchell, General Motors' styling czar, had hated that the first-gen Camaro had to be built on the plebian Chevy II bones, as it severely limited the amount of visual verve that could be styled into the car. Mitchell was an enthusiast of European sports cars, and he insisted that the second-generation Camaro would exude glamour and sportiness. And not just visually; the division wanted the Camaro to leapfrog the hated Mustang in performance. Chevrolet wanted the newest Camaro to be perceived as a virtually new car. When ads for the 1970 Camaro were released, they described the "new" car as "a sports car for the four of you." It was a touch of hyperbole, but just a small touch. The 1970

Camaro was arguably the finest-handling American car this side of the Corvette. And one of the most beautiful.

Chevrolet normally starts selling new cars in the autumn prior to the model year's release. This Camaro was different. The first 1970 Camaros didn't start showing up on dealer lots until February 26, 1970, while those same dealers still had 1969 models left to sell. Because of the late introduction, the first year of the second generation is often called 1970½ models. This late start had an effect on sales; the 1970 Camaro sold just over half as many units as the year before.

Like the first-generation Camaro, the second-gen used a unibody "core," with the front suspension and engine

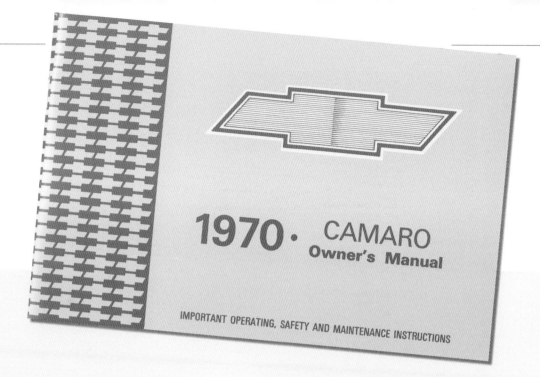

Wind tunnel research showed that the front chin and rear wing spoiler generated usable downforce, ideal for a vehicle with the performance potential of the 1970 Z28. The tall spoiler seen here was a COPO item. The attractive wheels were stamped steel, but looked like expensive forged units.

cradle bolted to the body using a subframe, similar to the first-generation Camaros. As the hard points for the 1970 Camaro were being determined, Engineering and Styling were at each other's throats. The engineers looked at the list of features that the new car was going to have, including air conditioning, and requested a design with a high cowl so that all of the mechanical bits would fit behind the dashboard. Styling would have none of that; Mitchell had requested a low cowl and a steeply raked windshield, and what Mitchell wanted, Mitchell generally got. When the dust had settled, the cowl was low, the windshield was dramatically raked, and the engineers spent many long hours getting the bits to fit.

Longer, lower, wider. For years, Detroit in general, and General Motors in particular, hewed to that mantra. People responded to a low-slung car in a way a tall, narrow vehicle could only dream about. While the 1970 Camaro's wheelbase was unchanged at 108.0 inches, the car was 2 inches longer, 1 inch lower, and almost 2 inches wider. Combined with curves that looked like they'd been lifted straight off a Ferrari Lusso, the newest Camaro readily stood out from the standard vehicular offerings. Toss in a wide range of powerplants, and the 1970 Camaro was a stellar leap forward in the pony car genre.

Most 1970 Camaros were fitted with a full-width front bumper, giving the gaping maw of a grille some protection.

Above: There was no small amount of anguish and worry about the vulnerability of the dramatic mile-per-hour barrier test, which showed that the pretty arrangement could not be used as the standard front end, but rules were bent to allow the stunning (and delicate) front end design to survive as part of the Rally Sport package.

Inset: Threading a line of cones was easy in the 1970 Z28. The suspension was tuned specifically for performance applications, and with a 1-inch front anti-roll bar and the F41 suspensions goodies, such as five-leaf rear springs, staggered rear shock absorbers, 125 lb-in rate rear springs, and a link-type rear stabilizer bar, it could carve cones with pure sports cars.

Right: For the first time in Camaro history, side-impact beams were built into the huge doors. The seats in the 1970 Z28 were comfortable, but provided virtually no lateral support. In a vehicle that handled as well as the Z28, the driver was challenged to remain behind the wheel during spirited driving.

Above: In the era before insurance companies dictated auto design, stylists could pen beautiful, flowing lines, as seen on the front of the 1970 Z28. The bumper was little better than a token; it was more a stylistic element than proper protection. The standard Z28 bumper was a full-width piece; the Rally Sport option used a pair of corner bumperettes.

Buyers that took home Rally Sport editions noticed that a pair of bumperettes graced the front corners of the cars, leaving the grille vulnerable to impact. But the visual impact was undeniable.

At the rear, a tall spoiler could be fitted, contributing effective downforce at speed. The story behind the entire aerodynamic package is worth telling. When Pontiac designer Bill Porter was researching aerodynamic data on the F-body, he came across some interesting info. Porter recalls, "Chevrolet engineering did a really extensive job testing the

Camaro in the wind tunnel. They really knew what that body would do. But they decided that they wouldn't use the data to help design the vehicle. One of the assistants at the wind tunnel, Doug Patterson, got me a copy of the test data, and I built the Trans Am slavishly off that data. I was like a kid in a candy store. The front spoilers, the side front spoilers that deflect air out over the wheels, the front spoiler is down to the ground as low as we could make it yet make ramp angles; we didn't really make the ramp angles, people knock the spoiler off all the time. The rear-deck spoiler was exactly

Above: One of Chevrolet's most popular engines was the LT1. In the 1970 Z28, the LT1 generated 370 lb-ft of torque and 360 SAE gross horsepower. With a solid lifter camshaft and plenty of forged components, it was a high-revving, race-bred powerplant.

Left: When the 1970 Z28 debuted, only one body style was available, the fastback coupe. Its lithe, flowing lines were a radical departure from the beefy design of the prior year. Goodyear Polyglas tires were standard on the Z28, and for the day, they were as good as it got. By contemporary standards, they were barely sufficient to hold the chassis off the ground, but they posted respectable numbers.

With its long hood, the engine compartment of the 1970 Z28 had plenty of room. With the engine pushed back as far as possible, the area in front of the radiator was considerable. In case of a frontal collision, that space helped protect the occupants, as the bumpers were little more than decorative strips of chrome.

to the position they [Chevrolet] found to be optimum." Once Chevrolet was aware how effective Pontiac's use of their data was, Chevy decided to use the information to fine-tune the aero profile of the newest Camaro.

Speaking of powerplants, a couple of changes were evident. With the latest Camaro having gained about 400 pounds, it was decided that the 140-horsepower, 230-cubic-inch straight-six just didn't have the guts to move the 1970 Camaro with anything resembling enthusiasm. Now the base engine was the mighty 250-cubic-inch six, with all of its 155 ponies straining to get out. Buyers who wanted a V-8 but didn't need a scorcher opted for the base 307-cubic-

inch engine. Rated at 200 horsepower, this understressed powerplant was topped with a two-barrel carburetor.

The 302-cubic-inch engine that had powered the Z/28 to such success was replaced by a 350-cubic-inch V-8. The SCCA had changed the rule regarding engine displacement, and now Chevy could run 350 cubic inches. Thus, the street car was fitted with the LT1 solid-lifter engine from the Corvette. With 11.0:1 compression, a forged crankshaft, four-bolt main bearing caps, and a 780-cfm Holley four-barrel carburetor, this high-revving screamer was rated at 360 horsepower. For the first time, the Z28 could be had with an automatic transmission as well as a four-speed manual. The beefy Turbo

Left: Unlike the broad-shouldered stance of the first-generation Camaro, the 1970 Z28 was a lithe, curvaceous vehicle in the European mold. This was not an accident; GM Styling Czar Bill Mitchell loved old-world automotive designs. He was a big fan of Ferrari, and the Italian exotic's influence is evident.

Below: The huge side window, called the Daylight Opening (DLO) in designer-speak, was devoid of a vent window, instead using GM's Astro Ventilation to purge the interior of odors. The elimination of a B-pillar gave the Camaro a sleek, flowing presence, but in a tight parking lot, the long door would prove to be a challenge.

Dual snorkels on the air cleaner assembly were an effort to reduce induction noise under heavy throttle. Automakers were becoming more aware of the need to lower the annoyance factor on their performance cars due to excessive sound. The air cleaner setup on the 1970 Z28 gave the occupants a pleasant sound level at steady cruising, yet under heavy throttle, the powerplant roared

1970

MODEL AVAILABILITY	two-door coupe
WHEELBASE	108 inches
LENGTH	188.0 inches
WIDTH	74.4 inches
HEIGHT	50.5 inches
WEIGHT	3,165 lbs
PRICE	$2,749
TRACK	61.3/60.0 inches (front/rear)
WHEELS	14 x 6 inches
TIRES	F70 x 14
CONSTRUCTION	unitized body/frame with bolt-on front subframe
SUSPENSION	long-arm/short-arm with coil springs front/longitudinal leaf springs, live axle rear
STEERING	recirculating ball
BRAKES	11-inch disc front, 9.5 x 2.0 inch drum rear
ENGINE	155-horsepower, 250-cubic-inch I-6; 200-horsepower, 307-cubic-inch V-8; 250- or 300- or 360-horsepower, 350-cubic-inch V-8; 350- or 375-horsepower, 402-cubic-inch V-8
BORE AND STROKE	3.88 x 3.53 inches (250), 3.87 x 3.25 inches (307), 4.00 x 3.48 inches (350), 4.13 x 3.76 inches (402)
COMPRESSION	8.5:1 (250), 9.0:1 (307, 250-horsepower 350), 10.25:1 (300-horsepower 350, 350-horsepower 402), 11.0:1 (360-horsepower 350, 375-horsepower 402)
FUEL DELIVERY	single-barrel (250), single two-barrel (307, 250-horsepower 350), single four-barrel (300- and 360-horsepower 350, 350- and 375-horsepower 402)
TRANSMISSION	three- and four-speed manual, two-speed automatic Powerglide, three-speed automatic Turbo Hydra-Matic
AXLE RATIO	Ranging from 2.73:1 to 4.10:1
PRODUCTION	12,578 6-cylinder, 112,323 V-8

Hydro-Matic 400 was the automatic used, a highly regarded gearbox. With the new body style, Chevrolet modified the logo of the Z28, eliminating the slash between the letter and numbers. Along with the capable RPO F41 sport suspension, and the heavy-duty springs, F60 tires, dual exhaust, and a lot more goodies, the Z28 option was $572.95 well spent. The hot shoes at *Hot Rod* magazine visited a drag strip with a Z28, going home with a 14.2-second time slip. With a lower center of gravity than the previous year's car, dedicated attention paid to the suspension, and improvements in tire technology, the 1970 Z28 could run rings around a 1969 version. Yet the rough edges that made the first-generation Z/28 so much fun to drive were polished to the point that the 1970 model was more Grand Tourer than street-legal race car.

Fans of big-block engines were pleased to see that the 396-cubic-inch V-8 continued to live under the long hood. A very slight increase in displacement meant that the engine packed 402 inches, but everyone called it the "396." The base version was rated at 350 horsepower in an attempt to deflect the stink-eye from the insurance companies, but with 415 lb-ft of torque, the same as the year before, few were fooled into thinking that this was anything but a tire destroyer. Buyers who could afford the insurance premiums, as well as having a stack of F70 x 14 replacement tires in the garage, could still order the big-block filled with 375 horsepower. Called RPO L78, it went for $385.50, and only 600 examples were built. Granted, it added 180 pounds over the front tires, but if you wanted to go quickly in a straight line, and didn't worry about compromised handling, the big-block was unforgettable. Speaking of quickly, the big-block would cover the quarter-mile in 15.3 seconds at 92.7 miles per hour. For a vehicle that tipped the scales at 3,850 pounds, on the tires of the day, that was an impressive feat.

Unlike the 1967 Camaro, which was a "get it out ASAP" vehicle, the 1970 Camaro design was started in 1967, giving Chevrolet plenty of time to style a cutting-edge automobile. The gestation period for automobiles in the 1960s was at least three years; today, a design production can be on the road within two years.

1971

For the second generation's second year, Chevrolet pretty much carried over the Camaro with few changes. A dozen new colors were introduced, along with fresh emblems. The front seats, which had been low-profile, were sourced from the Vega and higher seat backs, with headrests incorporated into the seat.

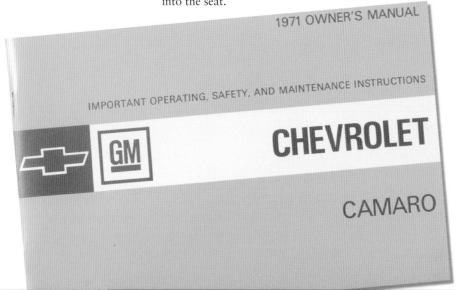

1971

MODEL AVAILABILITY	two-door coupe
WHEELBASE	108 inches
LENGTH	188.0 inches
WIDTH	74.4 inches
HEIGHT	50.5 inches
WEIGHT	3,208 lbs
PRICE	$2,758
TRACK	61.3/60.0 inches (front/rear)
WHEELS	14 x 6 inches
TIRES	E78 x 14
CONSTRUCTION	unitized body/frame with bolt-on front subframe
SUSPENSION	long-arm/short-arm with coil springs front/longitudinal leaf springs, live axle rear
STEERING	recirculating ball
BRAKES	11-inch disc front, 9.5 x 2.0 inch drum rear
ENGINE	110-horsepower, 250-cubic-inch I-6; 140-horsepower, 307-cubic-inch V-8; 165- or 210- or 275-horsepower, 350-cubic-inch V-8; or 260-horsepower, 402-cubic-inch V-8
BORE AND STROKE	3.88 x 3.53 inches (250), 3.87 x 3.25 inches (307), 4.00 x 3.48 inches (350), 4.13 x 3.76 inches (402)
COMPRESSION	8.5:1 (250, 307, 165- and 210-horsepower 350, 402), 9.0:1 (275-horsepower 350)
FUEL DELIVERY	single-barrel (250), single two-barrel (307, 165-horsepower 350), single four-barrel (210- and 275-horsepower 350, 402)
TRANSMISSION	three- and four-speed manual, two-speed automatic Powerglide, three-speed automatic Turbo Hydra-Matic
AXLE RATIO	Ranging from 2.73:1 to 4.10:1
PRODUCTION	11,178 six-cylinder, 103,452 V-8

One of the more controversial changes for 1971 was the adoption of net horsepower ratings from the gross ratings of years before. This resulted, on paper, in a perceived reduction in horsepower. Unfortunately, this perception was accurate, as General Motors had started lowering the compression in their engines in a move toward the widespread use of low-lead fuels. By replacing domed pistons with flat-topped pieces, the compression was effectively reduced for minimal cost.

GM was a bit hasty in the adoption of lower compression ratios, as virtually all of its competition didn't start lowering their engines' compression ratios until the release of the 1972 lineup. In the Camaro, that translated into fewer engines available on the order form. The 250-cubic-inch, 145-horsepower gross (110 net), straight six was still the base engine, and the entry-level V-8 displaced 307 cubic inches and was rated at 145 horsepower gross (110 net). Only two V-8s were on the option list. RPO L65 was a 350-cubic-inch powerplant that was rated at 245 horsepower, while RPO LS3 got a buyer a 396-cubic-inch mill advertised at 300 ponies. Customers who popped for the Z28 found that the 350-cubic-inch engine was now delivering 330 horsepower gross, down 30 from the year before. In literature, Chevrolet rated the Z28 at 275 horses. That diminishment of output was noticeable on the street and the track.

By looking at the thinned-out number of engine choices, it's clear that Chevrolet saw the writing on the wall regarding the future of muscle cars. Sales of 1971 Camaros were quite a bit lower than the preceding year, partly because of a United Auto Workers (UAW) strike from September 14, 1970, to November 22, 1970. Yet anyone who had driven a Camaro in the last couple of years could immediately tell that the government-mandated emission control devices, in conjunction with the reduced compression ratios, had a cooling effect on performance. This in turn had a cooling effect on sales.

All of the domestic carmakers in general, and Chevrolet in particular, were caught between a rock and a hard place. The government had mandated that ever-more stringent emission controls *had* to be on new vehicles, regardless of how the public might react to a significant diminishment in power. Chevy had to sell cars such as the Camaro with severely reduced outputs to satisfy the Feds, but the public still wanted vehicles that could get out of its own way.

So the public tended to look elsewhere. That didn't help GM's bottom line, so they couldn't devote enough resources to satisfy both the government and the public. They had to choose one group to please, and it had to be the government. It had the ability to shut the carmaker down. So the public was offered watered-down Camaros. Unsurprisingly, they responded just as anyone else would when presented with Grade 2 goods. They walked away.

Toward the end of the 1971 model year, dealers were slashing the prices on their remaining Camaros in an effort to clear the lot of the outgoing cars. Yet the 1971 Camaro was one heck of a GT car. So impressive was the Camaro that *Road & Track* magazine put it on their list of the 10 best vehicles for 1971. It's very telling that the Camaro was the only American vehicle on the list.

1972

For 1972, the Camaro offered more of the same, only less. Let me explain. Engine choices were the same as the year before, except for residents of California. The 396 was not available in the Golden State due to the expense of needing to certify it just for California. All of the other V-8 engines saw a decline in output as well, with the standard 307-cubic-inch mill dropping from 140 horsepower to 130. RPO L48, the 350-cubic-inch powerplant, saw a decrease in *oomph* go from 210 horses to 200. And the 49-state 396-cubic-inch engine fell from 260 to 240 horsepower net. The Z28, still boasting solid lifters within its 350 block, was now rated at 255 horsepower, down from 275. This resulted in a 0–60 miles per hour time of 7.5 seconds, with a quarter-mile result of 15.5 seconds at 90 miles per hour. The base six-cylinder engine still enjoyed a lofty 110 horsepower.

Appearance-wise, the 1972 Camaro is virtually indistinguishable from the 1971 model. Chevrolet stylists were readying the Camaro for the introduction of beefy bumpers for 1974. So the Camaro body was left alone for 1972, which, with its sleek lines and superb propositions, was not a bad thing. The interior was a carry-over as well, with the exception of a new inner door panel, complete with map pockets. Buyers noticed that the option list, usually as long as one's arm, had shrunk. Chevrolet was trying to control costs by reducing the number of available options that it had to keep in stock on the assembly line.

Even in base Sport Coupe form, the second-generation Camaro was still one of the most stylish cars being produced anywhere. *General Motors 2012*

The early 1970s were difficult times for the world's auto producers, but the freshness of the second-generation Camaro redesign kept sales relatively strong. *General Motors 2012*

The text on this ad for a hatchback Camaro was Greek for a reason—Chevrolet never built the car, leaving little reason to write actual advertising copy for it. *General Motors 2012*

While this image makes the rear seats appear comfortable, anyone who spent time in the back seat of a second-generation Camaro could attest that the seats were best used for storing small items. *General Motors 2012*

There's no such thing as a hard day's drive in a Camaro.

Whoever said you can't please everybody obviously hasn't seen our lineup of wagons and tailgates. Tailgates that go every way there is to go. Including A driving position that takes the tired out of a trip. A tailgate that disappears—the window up into the roof, the gate under the floor. A nice thing to have in tight places. (Our big Chevrolets have it.) A tailgate that folds down flat or swings open like a door. (Our mid-size Chevelles have it.) A tailgate that swings up and out of the way. (Our little Vegas have it.) A tailgate that swings up and down, or one that swings out like huge double doors. Take your choice. (Our Comfort is "baked in" the seats. Another tailgate that swings out like double doors is teamed up with a sliding side door. (Our versatile Sportvans have it.) Tailgates, of course, are not the end-all. Our wagons also give you five different load capacities. Plus ten different seating arrangements. We deliberately set out to please everybody. Because we want your new Chevrolet to be the best wagon you ever owned. Chevrolet

The Camaro's interior set new standards in sportiness for an American car. *General Motors 2012*

Late in the model year, the UAW didn't do the Camaro any favors, going on strike on April 7, 1972, shutting down the Norwood, Ohio, plant. At the time, this was the only facility building Camaros and Firebirds. The strike would last 117 days, and as the strike occurred toward the end of the model year, Chevrolet was in a bind when the strike was settled. Due to government regulations requiring impact-resistant bumpers and tougher emission controls starting in 1973, the 1,100 Camaros at the plant in various stages of assembly had to be scrapped. It was economically unfeasible to modify them to meet 1973 regulations, so they were crushed. When the dust settled, the 1972 Camaro would have the lowest build quantity until 1990, with just 68,651 units made. Some in Chevrolet felt that this was proof that interest in the Camaro was waning, and pressed for the plug being pulled on the F-body. Another group within General Motors had faith in the car, and pointed at the loyalty of the owners. Fortunately, the enthusiasts at GM prevailed, and the Camaro lived to fight another day.

1972 Chevrolet CAMARO OWNER'S MANUAL

IMPORTANT OPERATING, SAFETY AND MAINTENANCE INSTRUCTIONS

GM

1972

MODEL AVAILABILITY	two-door coupe
WHEELBASE	108 inches
LENGTH	188.0 inches
WIDTH	74.4 inches
HEIGHT	50.5 inches
WEIGHT	3,340 lbs
PRICE	$2,729
TRACK	61.3/60.0 inches (front/rear)
WHEELS	14 x 6 inches
TIRES	E78 x 14
CONSTRUCTION	unitized body/frame with bolt-on front subframe
SUSPENSION	long-arm/short-arm with coil springs front/longitudinal leaf springs, live axle rear
STEERING	recirculating ball
BRAKES	11-inch disc front, 9.5 x 2.0 inch drum rear
ENGINE	110-horsepower, 250-cubic-inch I-6; 130-horsepower, 307-cubic-inch V-8; 165- or 200-horsepower, 350-cubic-inch V-8; 240- or 255-horsepower, 402-cubic-inch V-8
BORE AND STROKE	3.88 x 3.53 inches (250), 3.87 x 3.25 inches (307), 4.00 x 3.48 inches (350), 4.13 x 3.76 inches (402)
COMPRESSION	8.5:1 (250, 307, 165- & 200-horsepower 350, 402), 9.0:1 (255-horsepower 350)
FUEL DELIVERY	single-barrel (250), single two-barrel (307, 165-horsepower 350), single four-barrel (200- and 255-horsepower 350, 402)
TRANSMISSION	three- and four-speed manual, two-speed automatic Powerglide, three-speed automatic Turbo Hydra-Matic
AXLE RATIO	Ranging from 2.73:1 to 4.10:1
PRODUCTION	4,821 six-cylinder, 63,651 V-8

Female customers have always comprised a large number of pony-car buyers, and Chevrolet did not neglect this demographic in its advertising. *General Motors 2012*

1973

Change was a-coming, and the first indication on the body of the Camaro was evident on the 1973 models. The bumpers were still chrome, but they now rested farther from the body, giving the beefed-up bumper support more room for deflection under impact. The Rally Sport version, with its gaping maw and bumperettes, was still in the lineup. But missing from the showroom was the Super Sport.

Chevrolet replaced the SS with a new edition, the Type LT, as the division felt that performance Camaros were a thing of the past. The marketing department was convinced that luxury was now preferred by customers, and the Type LT was the answer. Unlike the former SS option, the Type LT was a standalone model, not an RPO that could be added to other models. It was an option-heavy model, with the bulk of options aimed at comfort and convenience. Genuine simulated wood grain, dual rear view mirrors, and *lots* of

Government regulations were forcing changes on the Camaro by 1973, adding weight and robbing horsepower, but Chevrolet did an admirable job preserving the sleek lines of the Camaro, if not the overall performance. *General Motors 2012*

The year 1973 would be the last for the elegant chrome bumpers;

sound-deadening material was used throughout the cabin to impart a luxury ambience, and with a standard 350-cubic-inch V-8 that breathed through a dinky two-barrel carburetor, the resulting 145 horsepower was ideal for avoiding any hint of dreaded performance. The antique two-speed Powerglide was finally sent out to pasture. The Hydro Turbo-Matic three-speed automatic transmission now used a ratcheting Grand Haven shifter, allowing the driver to hold the gearbox in a gear until they deemed it appropriate to grab another cog.

Yet there were buyers who wanted both luxury and performance. They wanted it all. Chevrolet, never one to let a chance to sell multiple options pass it by, allowed the Type LT to be the base for a loaded vehicle. In fact, both the Rally Sport option and the Z28 option could be added to the Type LT. Thus you could order a Type LT Z28, a luxurious Grand Touring vehicle that packed a sporty powerplant. I say sporty in lower case due to the switch from solid valve lifters to hydraulic units for 1973. This prevented the Z28 engine from spinning so fast. In turn, Chevrolet could now install its excellent Harrison air conditioning on the Z28, an option that in years past had been unavailable due to high rpm's throwing off the drive belt.

1973

MODEL AVAILABILITY	two-door coupe
WHEELBASE	108 inches
LENGTH	188.5 inches
WIDTH	74.4 inches
HEIGHT	49.1 inches
WEIGHT	3,205 lbs
PRICE	$2,732
TRACK	61.3/60.0 inches (front/rear)
WHEELS	14 x 6 inches
TIRES	E78 x 14
CONSTRUCTION	unitized body/frame with bolt-on front subframe
SUSPENSION	long-arm/short-arm with coil springs front/longitudinal leaf springs, live axle rear
STEERING	recirculating ball
BRAKES	11-inch disc front, 9.5 x 2.0 inch drum rear
ENGINE	100-horsepower, 250-cubic-inch I-6; 115-horsepower, 307-cubic inch V-8; 145- or 175- or 245-horsepower, 350-cubic-inch V-8
BORE AND STROKE	3.88 x 3.53 inches (250), 3.87 x 3.25 inches (307), 4.00 x 3.48 inches (350)
COMPRESSION	8.25:1 (250, 307, 145- & 175-horsepower 350) 9.0:1 (245-horsepower 350)
FUEL DELIVERY	single-barrel (250), single two-barrel (307, 145-horsepower 350), single four-barrel (175- and 245-horsepower 350)
TRANSMISSION	three- and four-speed manual, three-speed automatic Turbo Hydra-Matic
AXLE RATIO	Ranging from 2.73:1 to 3.73:1
PRODUCTION	3,614 six-cylinder, 93,127 V-8

A 14-inch open-element air cleaner was fitted, and the copywriters at Chevrolet had fun with that, calling it "power on demand" sound. If it was only that easy. With hydraulic lifters now riding atop the camshaft, a cast-iron intake manifold replacing the aluminum part, and a Holley carburetor taking over for a Rochester Quadrajet, the 350-cubic-inch V-8 in the Z28 was now rated at 245 horsepower. Inside the block lived plenty of quality go-fast parts, including four-bolt main bearing caps holding a forged steel crankshaft, and 2.02/1.6-inch cylinder heads fitted with screw-in studs and guide plates. Pity the rest of the powerplant couldn't live up to its potential. The Z28 option cost $598.05, unless it was paired with a Type LT, then the performance RPO was only $502.05. Now that's value!

New to the Z28 was an air conditioning option. With the change to hydraulic lifters, the small-block wouldn't rev high enough to throw a belt (unlike the solid-lifter days) so A/C could be safely installed. This option helped make the Z28 Rally Sport a superb, all-around Grand Touring vehicle. Also new to the Camaro were power windows, as well as the widely hated seat belt buzzer. In late 1971, the federal government rolled back the excise tax; the upshot was that a 1973 Z28 cost *less* than a comparably equipped 1971 Z28. While overall brute performance was diminished, livability was improved, and with ever more stringent government regulations hitting the auto industry, the 1973 Camaro was the last year for the aggressive front end. RS models still used a pair of useless but great-looking bumperettes flanking the huge, gaping grille, but at least the structure behind the small bumpers was fitted with reinforcing bars that connected to the radiator core support, as well as reinforcing bars attaching the urethane nose to the core support, all in the name of crash resistance. The following year would see the introduction of mandatory 5-miles per hour front bumpers, and the graceful strips of chrome would be a memory. In 1974, a lot more of the Camaro's heritage would be driving off the order form.

1974

The most obvious change to the Camaro in 1974 was the massive, yet well integrated, front and rear bumper assemblies. Truly functional bumpers, in the governmental sense, were new to Detroit. The regulations required that the bumper handle a 5-miles per hour hit without deformation of the surrounding body structure, including headlights. The resulting fix at most manufacturers was to install a chrome-plated railroad tie onto the front of the car, with a pair of shock absorbers or flat leaf springs mounted behind the bumper to soak up the force of impact. Some automobiles in 1974 were more successful in incorporating the heavy

Left: In 1974, the United States was in a full-blown energy crisis, leading Chevrolet to put more emphasis on the Vega econocar than on the sporty Camaro. *General Motors 2012*

Below: While the safety bumpers of the 1974 Camaro weren't as hideous as they were on some of the Chevy's competitors, they still lacked the elegance of earlier designs. *General Motors 2012*

Left: The LT-1 engine in the Z28 still managed to generate 245 horsepower in 1974, a far cry from the 360 horsepower produced by the original in 1970, but loads better than the 185 horsepower available in the base 350 Sport Coupe. *General Motors 2012*

Right: In 1974, the original four-speed, LT-1–equipped Z28 made its last appearance. *General Motors 2012*

Below: The sleek lines at the rear of the Camaro would change after 1974; the following year, the rectangular rear window would give way to a Baroque curved window. *General Motors 2012*

The big bumpers required from 1974 on broke up the sleek lines of the original design, giving the car a droopy appearance. *General Motors 2012*

assembly into a vehicle's overall styling than others; the Camaro was among the cars whose front and rear ends were a design success.

The aluminum bumpers and the leaf springs added about 7 inches to the overall length of the vehicle, as well as 150 pounds. The base engine was a 350-cubic-inch V-8 rated at 145 horsepower, except in California, where it wasn't available at all. The engine hadn't been certified in the Golden State, so that state received RPO LM1, a 160-horsepower, 350-cube V-8. However, Californians *had* to pay an extra

$46 for the engine. If more power was wanted, the optional L48 350-cubic-inch engine ($76.00) was certified for sale in all 50 states, and it was advertised as belting out 185 ponies.

Yet even more thrust could be purchased, but only if you were willing to live with the Z28. The only engine available in the canyon-carver was the L82 V-8, rated at 245 horsepower. Unlike the 1973's open element air cleaner, the 1974 Z28 used a dual-snorkel air cleaner. This tended to muffle induction sounds under heavy throttle. Z28s built after January 1974 were fitted with the new High Energy

Left: Anything but subtle, the 1974 Camaro Z28 was increasingly dependant on tape graphics to generate excitement, as regulations and insurance had removed large amounts of underhood power. While the vivid stripes and logos made a strong visual statement, they also alerted the authorities that a vehicle with sporting intent was on the move.

Below: Standard under the long hood of the 1974 Z28 was the L82 350-cubic-inch V-8, rated at 245 horsepower. Aluminum valve covers and intake manifold were part of the package before engines disappeared beneath hoses and plastic covers. The snorkel air cleaner reduced induction noise and gave the car a subdued muscular sound.

Ignition (HEI) system in an effort to improve performance, fuel economy, and drivability. Prior to January 1974, the Z28 used the old-school points-style distributor. All this power would propel a Z28 down a drag strip in 15.2 seconds at 94.6 miles per hour. Its top speed was 123 miles per hour, but with the introduction of the national 55-miles per hour speed limit in 1974, it was clear that the government wanted everyone to slow down and preserve fuel and tires. Sure. Drivers who wanted the thrift of a six-cylinder engine had one engine on the roster, a 250-cubic-inch straight-six that pumped out 100 horsepower. In a 3,800-pound vehicle, the tires were safe from harm.

In an effort to keep the weight gain to a minimum, the front end was now built of fiberglass. Because the taillights now wrapped around into the rear fenders, new rear quarter panels were needed. More new items were on deck for 1974, including a widely detested seat belt interlock system that required a driver and passenger be buckled in before the vehicle would start. A heavy bag on the front passenger seat would trip the sensor, causing no small amount of frustration. Chevrolet responded to the outcry by removing the system in the middle of the model year.

Gone for 1974 was the Rally Sport. In the past, the Rally Sport version wore different front-end sheet metal and plastic than non-RS equipped Camaros. But with the introduction of the new front and rear end treatments, Chevrolet felt that the cost of developing special Rally Sport bumpers and surrounding bodywork would be too high. Thus the option was quietly omitted from the 1974 lineup.

1974 CHEVROLET

CAMARO
OWNER'S MANUAL

IMPORTANT OPERATING, SAFETY AND MAINTENANCE INSTRUCTIONS

Above: The new energy-absorbing aluminum bumpers on the 1974 Camaro line increased overall length by 7 inches. The 1974 Camaro Z28 is an overlooked collectible because it was the last year of true dual exhaust and still had a healthy 245 net horsepower from its L82 350-cubic-inch V-8.

Right: One of the primary tenets of classic pony car design is a small trunk, and the Camaro in 1979 embraced that requirement with gusto. The diminutive trunk was ideal for weekend trips, but for anything longer, it was likely that the rear seat got the overflow. It's not as if the back seats were holding people anyway.

1974

MODEL AVAILABILITY	two-door coupe
WHEELBASE	108 inches
LENGTH	195.4 inches
WIDTH	74.4 inches
HEIGHT	49.1 inches
WEIGHT	3,562 lbs
PRICE	$2,828
TRACK	61.3/60.0 inches (front/rear)
WHEELS	14 x 6 inches
TIRES	E78 x 14
CONSTRUCTION	unitized body/frame with bolt-on front subframe
SUSPENSION	long-arm/short-arm with coil springs front/longitudinal leaf springs, live axle rear
STEERING	recirculating ball
BRAKES	11-inch disc front, 9.5 x 2.0 inch drum rear
ENGINE	100-horsepower, 250-cubic-inch I-6; 145- or 160- or 185- or 245-horsepower V-8
BORE AND STROKE	3.88 x 3.53 inches (250), 400 x 3.48 inches (350)
COMPRESSION	8.25:1 (250), 8.5:1 (145-, 160-, 185-horsepower 350), 9.0:1 (245-horsepower 350)
FUEL DELIVERY	single-barrel (250), single two-barrel (145-horsepower 350), single four-barrel (160-, 185-, 245-horsepower 350)
TRANSMISSION	three- and four-speed manual, three-speed automatic Turbo Hydra-Matic
AXLE RATIO	Ranging from 2.73:1 to 3.55:1
PRODUCTION	22,210 six-cylinder, 128,798 V-8

Above: The stamped-steel mag-style wheels were a superb performance styling cue on the Z28 for the first half of the 1970s. During that time, they were painted a metallic-like argent color. When the Z28 returned to showrooms in 1977, the wheels were painted to match the body color.

One introduction on the option sheet new for 1974 was the availability of radial tires. Prices ranged from $104.15 for the blackwall version to $147.15 for the white-letter set. Sized at 14 inches, they would be the future rolling stock for virtually every car sold in America, but in 1974 they were an option. The Z28 was equipped with 15-inch tires, but they were bias-ply units.

Another mid year modification was the introduction of High Energy Ignition (HEI). Automobile ignition systems had long been the weak link in the pursuit of power, and with the introduction of HEI, Chevrolet could tailor the ignition system to work with current and future emission controls to give acceptable performance with unleaded fuel. The 1974 Camaro was the last year to use leaded gasoline; and with a new gas tank, the Camaro could now hold 21 gallons instead of 17. In January 1974, the Muncie four-speed was replaced by a Borg-Warner unit.

The Z28 got a real boost in visibility with its use in the International Race of Champions (IROC) series in 1974. Formed in 1973, the series pitted top-level drivers from a wide range of motorsports to a number of races in identically

Left: Chevrolet and Pontiac shared the rear spoiler center section on the Z28 and Trans Am, respectively. However the endcaps were unique to the body contours of each model. One of the challenges that the factory faced was getting an even gap, top to bottom, between the center section and endcaps.

prepared race cars. In 1973, the Porsche 911 RSR was the weapon of choice, but the costs were astronomical. In an effort to hold down expenses, as well as make the races more TV friendly, the Camaro Z28 was tapped for IROC duty in 1974. This would prove to be a boon to Z28 sales, as the old adage of "win on Sunday, sell on Monday" was alive and well in the form of 13,802 sold, making 1974 the second most successful year for the Z28 to date.

THERE IS NO LAW AGAINST DRIVING WITH A SMILE ON YOUR FACE. '75 CAMARO.

1975 highlights.

The Camaro Six is extensively refined, with a new cylinder head and new carburetion. You get greater fuel economy along with improved performance.

All 1975 Camaros are equipped with catalytic converters which help reduce hydrocarbon emissions by 56%, carbon monoxide by 61%.

Camaro's new High Energy Ignition system (HEI) delivers from 25% to 40% more usable voltage to the spark plugs for surer starts in all weather. No more points or condenser to replace at tune-up time.

The Camaro carburetor now inducts outside air, which is generally cooler and denser than hot under-hood air. The result is improved performance at all engine speeds, after warm-up.

Steel-belted radial ply tires are standard on all Camaros for 1975. There is less rolling resistance with radials, which increases gas mileage.

The use of cleaner unleaded fuel gives you more time between lubes, oil changes, filters and plugs. Your spark plugs should last at least 22,500 miles now.

New finned rear brake drums run cooler than the conventional kind, resisting fade.

You're driving slower now, and perhaps you're driving less. But nobody ever said you had to drive sad.
To that principle and all who applaud it, Chevrolet enthusiastically dedicates the 1975 Camaro.
A beautiful car.
To drive.
You can see from its shape that Camaro is a *driver's car.*

Low profile, wide stance, sloping hood and deck.
And the way it looks is the way it goes. Camaro responds eagerly when you touch the pedals or turn... pointed wher... takes a curve... of lean.
Camaro... Sensua... Camaro... been a com...

The wheelbase is 108 inches.
The standard engine in the standard model is a reasonably economical 250-cubic-inch Six, ... revamped for 1975.

Sport Coupe, LT or Z.
The important news, though, is that Camaro is what Camaro was: A sensibly sporty compact that looks like a million and drives like it looks.
There are two models: The ... Coupe and ...

enjoy it. All you have to do is drive it. Which we hope you'll do soon.
And with a smile.

CHEVROLET MAKES SENSE

Above: A change in the rear window shape for the 1975 model year gave the Camaro an entirely different character. *General Motors 2012*

Left: The Z28 disappeared for the 1975 model year, making the Type LT Sport Coupe the sportiest Camaro available. *General Motors 2012*

Opposite top: The Camaro still looked sporty in 1975, but with a maximum power rating of 155 horsepower, it didn't have the snort to back up its good looks. *General Motors 2012*

Opposite bottom: With the death of the Z-28, the RS appearance package took up the outrageous tape graphic banner. *General Motors 2012*

1975

For 1975, Chevrolet giveth, and Chevrolet taketh away. The division returned the Rally Sport to the mix, yet removed the Z28 from the showroom. In the matter of the Z28, it was clear to Chevrolet that the mandated regulations would essentially emasculate the car. So rather than let it become a laughing-stock, they wisely pulled it. This appears to be a last-minute decision on Chevrolet's part. Early sales materials list the car

as available, and Chevy's records indicate that one 1975 Z28 was built. What isn't known is if it went to a customer or was a prototype. If it's still around, it might be worth a tidy sum.

On the Rally Sport front, the 1975 RS wore the same sheet metal as every other Camaro. But for $238 (or $165 when ordered with the LT package) you took home a whole lot of visual pizzazz and not much else. The Rally Sport option, RPO Z85, was a paint and tape package, nothing

Above: Sales suffered almost as much as performance in 1975, but the Camaro still sold a respectable 145,770 units. *General Motors 2012*

Right: While its 155 horses wouldn't generate any serious velocity, the Camaro certainly looked quick, especially with the RS appearance package. *General Motors 2012*

more. True, the rear window was now a wraparound design, in an attempt to reduce the huge blind spot behind the C-pillar. But every 1975 Camaro got that.

No, performance was a memory. The emphasis was now firmly in the luxury camp. With bird-s-eye maple wood grain spread across the dashboard, deep pile carpet underfoot, power door locks, and more choices of radios than V-8 engines, it was clear that the good life was measured in decibels and not miles per hour. The most powerful engine available in the 1975 Camaro was a 350-cubic-inch V-8 rated at just 155 horsepower and 250 lb-ft of torque. A healthy kid on a 10-speed could threaten it. Heaven forbid a six-cylinder mill reside beneath the long hood when you tried to merge onto a freeway.

Unleaded fuel became the norm in 1975, and that required a number of engineering modifications, meaning that Camaros now utilized a catalytic converter in the exhaust system, an "unleaded fuel only" decal at the filler, a restrictor filler neck that would not allow a leaded-fuel nozzle in, and a screw-on fuel filler cap. This last component was used to ensure the fuel system was a closed system, minimizing the amount of fuel vapors to escape into the atmosphere. To accommodate the required catalytic converter, a bulge was built into the front passenger floor. This protrusion would be a constant for the rest of the Camaro's life, and on warm days the heat from the converter would soak through the floor and carpet to toast passengers' legs.

Each year, government regulations piled ever more stringent fuel emission standards on the backs of the automakers, whether the technology was ready or not. These Draconian measures resulted in manufacturers essentially emasculating their powertrains with primitive emission control devices to meet the letter of the law. Eventually, computers would play an increasing role in engine management. But until then, American automobiles were lucky to be able to generate enough power to get out of their own way.

The editors of *Car and Driver* magazine sauntered the length of a drag strip in a 155-horse V-8, taking a leisurely 16.8 seconds. Never would a magazine test as slow a Camaro. After this showing, the Camaro would only get faster. Incrementally, but still improving. Yet with total sales of 145,770 for model year 1975, it was clear that the product planners at Chevrolet were on their game for their

1975	
MODEL AVAILABILITY	two-door coupe
WHEELBASE	108 inches
LENGTH	195.4 inches
WIDTH	74.4 inches
HEIGHT	49.1 inches
WEIGHT	3,723 lbs
PRICE	$3,553
TRACK	61.3/60.0 inches (front/rear)
WHEELS	14 x 6 inches
TIRES	E78 x 14
CONSTRUCTION	unitized body/frame with bolt-on front subframe
SUSPENSION	long-arm/short-arm with coil springs front/longitudinal leaf springs, live axle rear
STEERING	recirculating ball
BRAKES	11-inch disc front, 9.5 x 2.0 inch drum rear
ENGINE	105-horsepower, 250-cubic-inch I-6; 145-, 155-horsepower, 350-cubic-inch V-8
BORE AND STROKE	3.88 x 3.53 inches (250), 4.99 x 3.48 inches (350)
COMPRESSION	8.25:1(250), 8.5:1 (350)
FUEL DELIVERY	single-barrel (250), single two-barrel (145-horsepower 350), single four-barrel (155-horsepower 350)
TRANSMISSION	three- and four-speed manual, three-speed automatic Turbo Hydra-Matic
AXLE RATIO	2.73:1, 3.08:1
PRODUCTION	29,749 six-cylinder, 145,770 V-8

intended buyers. Good styling sold and continues to sell cars. The Dodge Challenger and Plymouth Barracuda ceased production at the end of the 1974 model year, so sporty car buyers in 1975 could choose between the Camaro and the pathetic Mustang II. Enthusiasts shied away from the little Ford like the plague, so the Camaro was essentially playing in a field of one. Chevrolet's marketing staff was very happy.

Above: While performance had decreased, the Camaro was still an infinitely more desirable car than the universally reviled Pinto-based Mustang that Ford introduced in 1975. The Ford handily outsold the Camaro, which was unfortunate for both Chevrolet and the poor sots who bought the miserable Mustang. *General Motors 2012*

1976

As America celebrated its bicentennial, the Camaro was pretty much a carry-over from 1975. A few bits of exterior brightwork kept the keen-eyed busy, cruise control (Cruise Master) was introduced and, due to demand, production was expanded from the Norwood, Ohio, plant to include Van Nuys, California, for the first time since 1971. Power was up slightly. The top-rated V-8, RPO LM1, displaced 350 cubic inches and now cranked out 165 horsepower. The base V-8, a 305-cubic-inch iron block, generated 140 horses. Under the hood, that was it. But the Rally Sport was not a car that could be overlooked, at least from a visual standpoint. A low-gloss black finish was applied to the hood, tops of the front fenders, headlight bezels, grille, upper portion of the doors and side glass, and rear valiance. A tri-color band separated the black portions from the body color, and Rally Sport

decals finished the graphics package. While none of these items made the vehicle any faster, it was a serious dollop of visual pizzazz, and in the mid-1970s that was the closest most could get to performance.

Yet the 1976 Camaro posted the highest sales numbers yet for the second-generation Camaro. V-8-equipped Camaros totaled 144,912, while the six-cylinder-equipped version sold 38,047 units. That equals 182,959—an impressive number in anyone's book. It was clear that the public loved the Camaro, and Chevrolet wanted the faithful to continue to frequent Bowtie showrooms. But even Chevy knew that they could only install so many tape and paint packages, at which time the buyers would realize that while the Camaro had plenty of show, it was painfully short of go. So engineers were burning the midnight oil in an effort to put a few more beans beneath the long hood.

In hindsight, it's easy to criticize the performance of mid-1970s Camaros, but this criticism applies to every car produced during that period, in America or anywhere else. *General Motors 2012*

Above: After a one-year hiatus, the Z28 returned for duty in 1976. *General Motors 2012*

1976 Nova Camaro Owner's Manual

1976 Chevrolet Important operating, safety and maintenance instructions

1976

MODEL AVAILABILITY	two-door coupe
WHEELBASE	108 inches
LENGTH	195.4 inches
WIDTH	74.4 inches
HEIGHT	49.1 inches
WEIGHT	3,679 lbs
PRICE	$3,762
TRACK	61.3/60.0 inches (front/rear)
WHEELS	14 x 6 inches
TIRES	E78 x 14
CONSTRUCTION	unitized body/frame with bolt-on front subframe
SUSPENSION	long-arm/short-arm with coil springs front/longitudinal leaf springs, live axle rear
STEERING	recirculating ball
BRAKES	11-inch disc front, 9.5 x 2.0 inch drum rear
ENGINE	105-horsepower, 250-cubic-inch I-6; 140-horsepower, 305-cubic-inch V-8; 165-horsepower, 350-cubic-inch V-8
BORE AND STROKE	3.88 x 3.53 inches (250), 4.00 x 3.48 inches (350)
COMPRESSION	8.25:1 (250), 8.5:1 (350)
FUEL DELIVERY	single-barrel, (250), single two-barrel (145-horsepower 350), single four-barrel (155-horsepower 350)
TRANSMISSION	three- and four-speed manual, three-speed automatic Turbo Hydra-Matic
AXLE RATIO	2.73:1, 3.08:1
PRODUCTION	38,047 six-cylinder, 144,912 V-8

1977

Things in CamaroLand improved for 1977. Deep within the bowels of Chevrolet, gearheads had been toiling away, determined to put a performance Camaro back on the road. Midway through the model year, the Z28 returned! Thanks for this development has to go to sister division Pontiac. In the mid-1970s, when the power output of the Camaro fell to near catatonic levels, Pontiac was stuffing serious power (for the times) under the hood of their Trans Am. For instance, in 1975, when the Camaro's biggest engine made just 155 horsepower, the 455-cubic-inch Pontiac generated 200 horses, and an impressive 330 lb-ft of torque.

In 1977, the huge 455 of Pontiac was replaced by a 400-cubic-inch engine that put out the same 200 horsepower, and just 5 lb-ft less torque. In the days before brand marketing, when the divisions actually had vehicles specific to that division, and the divisions hated each other, Chevrolet was major league mad that Pontiac was kicking it in the performance tail. Chevy felt, with good reason, that they were losing a lot of performance-oriented customers to Pontiac. So the Bowtie staff was working overtime to get the Camaro back in the performance game. The result of their efforts was in the showrooms in the middle of the model year: the return of the Z28. Unlike prior years, the 1977 Z28 was a standalone model, and it was not possible to combine it with the Rally Sport or Type LT models.

Knowing that power increases were going to be incremental and small, Chevrolet engineers devoted most of their time bringing the suspension up to world-class levels, while the designers created a bold visual package that made the Z28 stand out from the crowd. Under the hood was the reliable 350-cubic-inch engine, rated at 185 horsepower in 49 states, 175 in Hollywood. Torque, so important on the street, dialed in at 280 lb-ft. The Z28 was available with both a four-speed Borg-Warner Super T-10 manual and a three-speed Turbo Hydra-Matic automatic transmission, unless you lived in California. Only the auto was offered there.

It was in the arena of handling that the 1977 Z28 really shined. Chevrolet had invested considerable resources to create a broad-shouldered bruiser that could corner with cars costing far more, usually from Europe. Boasting dual exhaust resonators, the Z28 had the traditional American rumble and snarl so prized by enthusiasts. With sway bars at both

1977	
MODEL AVAILABILITY	two-door coupe
WHEELBASE	108 inches
LENGTH	195.4 inches
WIDTH	74.4 inches
HEIGHT	49.1 inches
WEIGHT	3,828 lbs
PRICE	$4,113
TRACK	61.3/60.0 inches (front/rear)
WHEELS	14 x 6 inches
TIRES	E78 x 14
CONSTRUCTION	unitized body/frame with bolt-on front subframe
SUSPENSION	long-arm/short-arm with coil springs front/longitudinal leaf springs, live axle rear
STEERING	recirculating ball
BRAKES	11-inch disc front, 9.5 x 2.0 inch drum rear
ENGINE	90-, 110-horsepower 250-cubic-inch I-6; 135-, 145-horsepower 305-cubic-inch V-8; 160-, 170-, 185-horsepower, 350-cubic-inch V-8
BORE AND STROKE	3.88 x 3.53 inches (250), 3.74 x 3.48 inches (305), 4.00 x 3.48 inches (350)
COMPRESSION	8.3:1 (250), 8.4:1 (305), 8.2:1 (350)
FUEL DELIVERY	single-barrel, (250), single two-barrel (145-horsepower 350), single four-barrel (155-horsepower 350)
TRANSMISSION	three- and four-speed manual, three-speed automatic Turbo Hydra-Matic
AXLE RATIO	2.56:1, 2.73:1, 3.08:1
PRODUCTION	31,389 six-cylinder, 218,853 V-8

ends, firmer springs, tuned shock absorbers, and quicker 14:1 steering, Chevy had brought the handling of the Z28 into a different level than years before. Radial tires were standard, and Chevy engineers had gotten a handle in efforts to tune the suspension to extract maximum performance from the rolling stock. In states other than California, the small-block was bolted to a Borg-Warner M21 close-ratio manual gearbox.

The exterior of the Z28 was treated to the full graphics treatment. With a blacked-out grille, spoilers front and

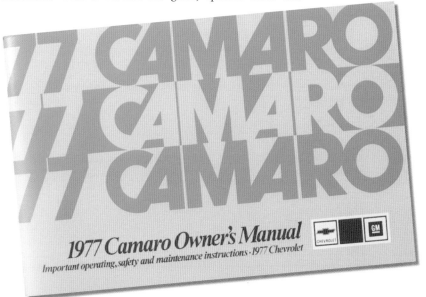

1977 Camaro Owner's Manual
Important operating, safety and maintenance instructions · 1977 Chevrolet

Right: With a wraparound instrument panel, well-placed shifter, sport steering wheel, and high-back bucket seats, the interior of a 1977 Camaro wasn't a bad place to spend time. While white seats may have been challenging to keep clean, they gave the Camaro an upscale appearance.

Below: Not every Camaro built in 1977 was a V-8-powered road warrior. The standard Type LT was equipped with at 250-cubic-inch inline six-cylinder engine delivering a lusty 110 horsepower (90 in California).

rear, body-colored wheels and bumpers, and tape and trim enough to cover a small house, the Z28 flew anywhere but under the radar. On the drag strip, it posted an improved performance from the dismal 1975 model year: 15.4 seconds. Most drivers got closer to 17 seconds. Part of the faster time was due to *Hot Rod* and *Motor Trend* magazine's abusive shift technique, grabbing gears sans clutch. But a large part was due to the strong engine. With the four-barrel carburetor and dual exhaust flowing efficiently, the Z28 made about 30 horsepower more than the stated level. Because of the expense of re-certifying an engine with a new exhaust, Chevrolet kept the increased power output under wraps and let the car do the talking out on the strip and street. The customers talked with their wallets, and the 1977 Z28 racked up 14,347 sales, impressive for a mid year release.

The rest of the 1977 Camaro line benefited from the halo effect of the Z28. Granted, engine choices were a bit sparse, and power increases were small. But they *were* increases, not an ebbing of output. Soldiering on was the 250-cubic-inch

straight six-cylinder engine, now the standard mill, churning out 110 horsepower with its single-barrel carb, unless the buyer lived in California. Then the trusty powerplant made a whopping 90 horses.

All V-8s were optional in 1977, unless you ordered the Z28; it only came with the beefy 350. The 305-cubic-inch engine was finding its way into more Camaros, as buyers wanted more grunt than the 250-cubic-inch six offered, but didn't want to pay out for the optional 350 mill, as well as the increased insurance premiums that the larger engine required. Ordering the two-barrel-equipped, $120 RPO LG3 would put the 145-horsepower (135 in California) 305 in the engine compartment. It was a popular choice in 1977, as 147,173 Camaros were so equipped. Need a little more *oomph*? Then the 350-cubic-inch V-8 was your next stop. Known as RPO LM1, this $210 option delivered 170 horsepower (160 in California), and 270 lb-ft of torque, thanks to its four-barrel carburetor. More than 40,000 customers sprang for this engine.

Camaro Type LT buyers tended to be a bit more mainstream, eschewing flashy graphics. For 1977, Chevrolet curried favor with these buyers by offering tasteful striping, cross-lace alloy wheels, and whitewall tires. The suspension leaned toward comfort rather than pure performance.

By the time the books closed on the 1977 Camaro, it had set a new record; for the first time in the Camaro's history, it outsold the Ford Mustang. Granted, the Ford was the horrid little Mustang II. But for many years, the Blue Oval had bested the Bowtie in pony car sales. Score: Camaro 198,755, Mustang 161,654. Finally, Chevy owned best-selling bragging rights. And now that it had topped the Mustang, Chevrolet wanted to keep the momentum up.

1978

It did it by releasing a restyled Camaro. In any vehicle's life, a manufacturer will give a model a midcycle freshening, keeping the public's interest alive while developing the next generation. Often, this freshening consists of revised front and rear sheet metal, new interior materials, and a brace of new colors and options. This was the path the Camaro followed for the 1978 model year.

Slippery front and rear body color urethane bumper covers gave the Camaro a much more integrated look, unlike years prior when the bumpers resembled chrome-plated railroad ties bolted to the vehicle. Now the design looked far more flowing. And while most of the powertrains were carried

The 1978 Camaro Type LT was aimed directly at the Camaro buyer who appreciated sporty good looks and a rewarding driving experience, but didn't want the attention that the Z28 model engendered, especially from law enforcement. A plush, comfortable cruiser, it was an ideal cross-county conveyance.

Most 1978 Camaro Type LTs were equipped with three-speed automatic transmissions. The four-speed manual was something of a rarity in a vehicle designed for effortless cruising. To engage reverse gear, the shifter had to be lifted slightly, then slid into gear.

over from 1977, the overall package was more attractive. There were five different Camaros available, ranging from the economical Sport Coupe equipped with a base 250-cubic-inch L-6, to the beefy Z28, recipient of a 350-cubic-inch V-8. Between them were a Type LT Sport Coupe, Rally Sport Coupe, and the Type LT Rally Sport Coupe. Confusing? You bet. But buyers flocked to showrooms. When the books were closed on the 1978 model year, a new record had been set, with 272, 631 Camaros having been sold. Let's see what all the fuss was about.

Under the hood, little had changed. The top dog Z28 was rated at 185 horsepower, 175 in California. This cast-iron powerplant used a hydraulic camshaft, 8.2:1 compression, two-bolt main bearings, and a Rochester four-barrel carburetor. The exhaust system didn't use a muffler, instead depending on a single catalytic converter and a pair of resonators to give the dual exhaust tips a throaty growl. The Golden State was also deprived of a manual transmission.

Yet driving the Z28 returned a quality experience. Chevrolet had made subtle upgrades to the suspension, adding lower control arm reinforcements and beefing up the rear spring shackles. Aluminum wheels, a $265 option, were available for the first time on the Z28, and they reduced unsprung weight, which in turn improved ride quality and handling. While raw performance was not to be found beneath the hood, the Z28 made a strong visual statement. With its blacked-out grille, non-functional hood scoop, and front fender vents, the Z28 looked like a race car that stumbled onto the street. Evidently, a lot of customers wanted to be mistaken for race car drivers, as 54,907 were sold. With a new "string"-wrapped steering wheel, lush interior materials, and Z28 graphics scattered throughout the cockpit, the top-range Camaro was a comfortable place to be.

Throughout the Camaro line, the emphasis was on creature comforts with a dose of performance. Chevrolet engineers were burning the midnight oil trying to regain grunt lost to federally mandated emission regulations, as well as the weight gain due to required safety equipment. For 1978, the Camaro Z28 tipped the scales at 3,560 pounds, yet Engineering fitted the all Camaros except the Z28 with numerically lower rear axle ratios in an attempt to improve fuel economy. For Z28s equipped with a Turbo Hydra-Matic 350 automatic transmission, a set of 3.42:1 gears

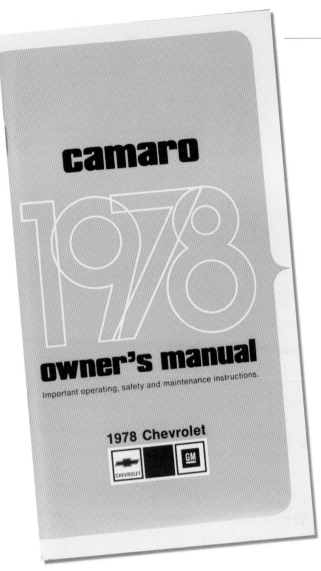

camaro
1978

owner's manual

Important operating, safety and maintenance instructions.

1978 Chevrolet

1978	
MODEL AVAILABILITY	two-door coupe
WHEELBASE	108 inches
LENGTH	195.5 inches
WIDTH	74.4 inches
HEIGHT	49.2 inches
WEIGHT	3,616 lbs
PRICE	$4,414
TRACK	61.3/60.0 inches (front/rear)
WHEELS	14 x 6 inches
TIRES	E78 x 14
CONSTRUCTION	unitized body/frame with bolt-on front subframe
SUSPENSION	long-arm/short-arm with coil springs front/longitudinal leaf springs, live axle rear
STEERING	recirculating ball
BRAKES	11-inch disc front, 9.5 x 2.0 inch drum rear
ENGINE	90- & 110-horsepower (250); 135-, 145-horsepower (305); 160-, 170-, 175-, 185-horsepower (350)
BORE AND STROKE	3.875 x 3.53 (250), 3.736 x 3.48 (305), 4.00 x 3.48 (350)
COMPRESSION	8.1:1 (250), 8.4:1 (305), 8.2:1 (350)
FUEL DELIVERY	single-barrel (250), single two-barrel (305), single four-barrel (350)
TRANSMISSION	three- and four-speed manual, three-speed automatic Turbo Hydra-Matic
AXLE RATIO	ranging from 2.73:1 to 3.42:1
PRODUCTION	36,982 six-cylinder, 235,649 V-8

At normal highway speeds, the tall rear spoiler on a 1978 Camaro didn't generate much effective downforce, only coming into its own at go-directly-to-jail speeds. But nobody could deny that the spoiler added considerable visual excitement to Chevrolet's F-body.

were used; Borg-Warner T-10 manual four-speed cars used a 3.73:1 ratio. The result by *Car and Driver* magazine on the drag strip was good for the day, with a 16.0-second run at 91.1-miles per hour trap speed. Top speed was 123 miles per hour, and bringing the Z28 to a stop from 70 miles per hour took 181 feet. For the day, this was impressive performance. In fact, the 1978 Camaro was the top-selling American car in West Germany, a country that knows a thing or two about performance.

The rest of the Camaro line basked in the glamour of the Z28, and while they weren't the serious road warriors that the Z28 was, they could be optioned to provide an interesting drive. The Rally Sport was a separate Camaro model, not an RPO, and while it came with the base 110-horsepower (90 in California) 250-cubic-inch straight-six cylinder engine, it only took $185 to upgrade to the 145-horsepower (135 in California) two-barrel, 305-cubic-inch V-8. Buyers wanting a bit more beans could spend $300 and take home RPO LM-1, the non-Z28, four-barrel, 350-cubic-inch V-8 that was rated

at 170 horses (160 in California). Both optional V-8s were popular, as 143,110 of the 305-cubic-inch engine were sold, while 92,539 buyers sprang for the 350 engine. It was clear that the public was hungry for performance. For the second year, the Camaro handily outsold arch-rival Mustang; 247,437 units sold to the little Ford's 179,039. It wasn't even close.

Partway through the model year, a new option appeared, RPO CC1, removable glass roof panels. This $625 option was

Above: The convertible had disappeared from the Camaro lineup in 1970, but buyers wanting fresh air and blue sky could soon order the optional T-top roof treatment. Listed as RPO CC1, this was a $625 option, and 9,875 buyers checked the box on the order form.

Right: The Type LT was the top-selling Camaro model in 1978, with 71,331 built. Available with either a six- or an eight-cylinder engine, it stressed comfort and luxury in a sporty-looking package. The base price for a six-pot Type LT was $4,814, but with an option list as long as your arm, it didn't take long to fatten the bottom line.

created to mollify the part of the public that was clamoring for a convertible. This option was a reasonable alternative to a proper convertible, giving occupants a fresh-air experience while maintaining vehicle security. This pricey option was popular, with 9,875 sets sold. The Camaro was riding high, and the two-millionth Camaro was built at the Van Nuys, California, plant on May 11, 1978.

1979

Chevrolet was encouraged by the public's response to the 1978 Camaro, and while changes for 1979 weren't major, they kept the Camaro in the lead for overall segment sales. The most significant change in the Camaro lineup was the dropping of the two LT models, replaced by a single Berlinetta model. Once again, the Rally Sport was a separate model, and the Z28 topped out the range. With the exterior changes that debuted in 1978, Chevrolet wasn't going to be making any big changes for 1979. Minor trim changes to all Camaro models, and a three-piece chin spoiler for the Z28, were the extent of exterior modifications, but the interior enjoyed a new dashboard and instrument panel. Evidence that comfort and convenience played a major role in keeping customers happy were six sound systems on the option list, including two with CB radios.

Under the hood, things were looking up for non-Z28s— slightly, if you lived in California. While the same slate

Above: The 1979 Camaro Z28 was flush with aerodynamic aids, including a deep chin spoiler, NACA hood scoop, and tall rear spoiler. No doubt these aided in cutting the wind around the Camaro, but at street-legal speeds, their effectiveness was minimal. But like so much in the 1970s, style over substance was the norm.

Above: The fender vents were functional, allowing hot air from under the hood to exhaust to atmosphere. Stamped-steel wheels with trim rings had long been a design element on various Chevrolet vehicles, and the 1979 Z28 wore the package well. An embossed Z28 center cap kept the dirt out of the bearings.

Right: Beneath the large black air cleaner assembly lived a Rochester Quadrajet four-barrel carburetor atop the LM1 350-cubic-inch V-8. The 49-state version delivered 175 horsepower, while California versions churned out 170 ponies. Buyers in California had only one transmission choice, the three-speed Turbo Hydra-Matic automatic.

Below: In a vehicle like the 1979 Camaro Z28, large doors to allow access to the rear seats were necessary. The sheer size of the doors could pose a challenge in a tight parking lot, while the weight of the door, a large metal structure with a side impact beam inside, could prove a handful if the car was parked on a grade.

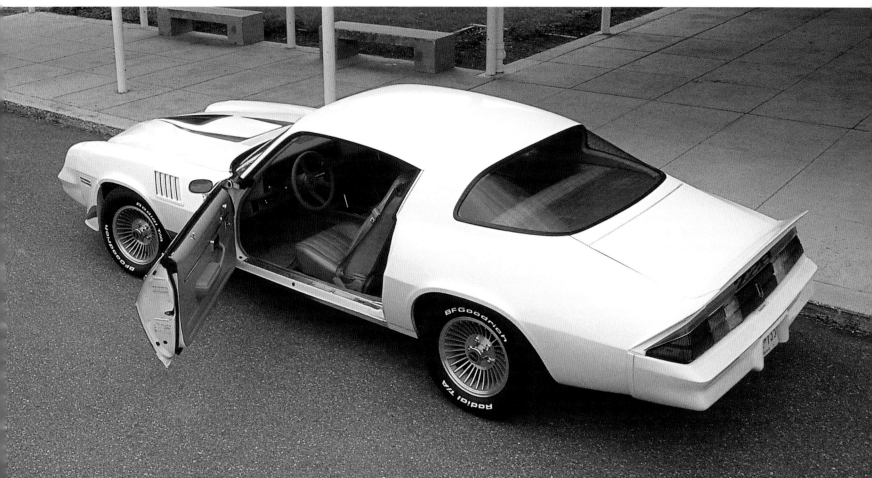

of engines used in 1978 were on the menu, powerplants bound for California received a massive power infusion of 5 horsepower. Beats a cut in output. However, the 1979 Z28 actually had a power reduction. The induction and exhaust systems were modified, along with ignition timing. The result was a small drop to 175 horsepower in 49 States, while California made do with 170 ponies. Once again, a manual transmission was not available in California. All 50 states enjoyed the same excellent Z28 suspension, including 11-inch front disc brakes rotors, 1.2-inch front and 0.55-inch rear stabilizer bars, and Goodyear GR70-15 steel-belted radial tires. The performance Camaro was an excellent seller, with 84,887 rolling out of showrooms.

All V-8-equipped Camaros used a numerical lower rear axle ratio gear set in an effort to improve fuel economy. The fuel crisis from just a couple of years prior had spooked Detroit, and it was determined to provide vehicles with good economy, regardless of the effect on performance. Fortunately, Chevrolet's engineering staff was learning how to increase power levels without losing fuel economy and

Right: The tail spoiler on the second-generation Camaro was a result of extensive wind tunnel development work. Pontiac's designers, notably Bill Porter, "borrowed" the data Chevrolet generated in the wind tunnel and created the distinctive aero package used on the Trans Am. At the time, both Chevy and Pontiac were competing in the Trans-Am series, and any edge on the track was seized.

Above: A decade and a half after the Camaro hit the road, the 1979 model continued exhibiting the graceful lines that characterized the model since its introduction. With a long hood, short deck, and sloping rear window, it was a timeless blend of sports and personal car.

Left: The stamped-steel 15 x 7 wheels on the 1979 Camaro Z28 wore P225 x 70R15 radial tires. Inside the front wheels were power disc brakes, with drum brakes at the rear, standard on the Z28. Chevrolet was very aware that slowing a sporty car was as important as accelerating it.

Above: This Camaro Z28 was one of 11,875 Camaros clad in this finish, paint code 51, Bright Yellow, in 1979. The most popular color on the Camaro that year was Dark Blue, while the rarest hue was Light Green. For low-key driving, Bright Yellow wasn't an ideal choice.

Right: With dusk approaching and a wide-open road beckoning, the driver of this 1979 Z28 could look forward to an evening of memorable cruising. With the huge rear window, the view over the driver's shoulder was better than many other vehicles of the era.

Above: Emission controls were in their infancy, and the result was a Gordian knot of hoses, wires, and components covering the engine and filling every nook and cranny. With computers just starting to be used in automotive applications, and an old-school carburetor on top of the mill, reaching emission goals while maintaining drivability was an engineering challenge of the first magnitude.

Right: Not for the shy, the 1979 Camaro Z28 used a blacked-out treatment on the grille, headlight buckets, and parking light bezels, and a large decal surrounding the nonfunctional hood scoop. The Camaro had its best sales year ever in 1979, with Chevrolet selling a whopping 282,582 units.

Below: When the second-generation Camaro was on the drawing boards, the designers had no idea that the front and rear of the vehicle would have to be re-engineered and re-styled to accommodate a 5-miles-per-hour bumper on the front and a 2.5-miles-per-hour bumper at the stern. Yet the result was graceful as well as effective.

Above: If bold visuals determined how fast a vehicle was, the 1979 Camaro Z28 would break the sound barrier. With its handsome body-color stamped-steel wheels, air dam and spoiler, tape graphics, and subtle bulges, the Z28 sure looked the part. Unfortunately, with just 175 horsepower in 49-state guise, and 170 in California, the tires were safe.

Left: As in prior years, the 1979 Camaro Z28 enjoyed a full slate of performance enhancements, including a close-ratio four-speed manual transmission, dual exhaust, special springs and shock absorbers, and a 350-cubic-inch V-8 rated at 175 horsepower (170 in California).

Right: Unleaded gasoline was the only fuel allowed to power the 1979 Camaro Z28, preferably premium grade. The need for automotive engines to generate fewer pollutants put engineers on a crash course developing efficient emission controls. While effective, the power output of the engine suffered greatly.

Right: Chevrolet introduced a wraparound front chin spoiler on the 1979 Z28 in an effort to improve aerodynamics and fuel economy, as well as impart a "race car" feel to the styling. The louvers on the front fenders were functional, venting hot air from the engine compartment.

compromising emission controls. This was a struggle that all of the manufacturers faced, and as years went by, the performance slowly returned.

For Camaro buyers with a hankering for luxury over performance, the Berlinetta fit the bill with its standard Custom interior, while the exterior was awash in bright bits to the grille, glass trim molding, pinstriping, body-color sport mirrors, special argent appliquè to the area between the taillights, and special wheels. In an effort to make the interior as quiet as possible, Amberlite insulation blankets were used beneath the carpets, under the package tray, in the doors and rear quarters, in the roof and C-pillars, under the hood, and behind the rear seat. Dual horns were standard as well, though it's doubtful that occupants could actually hear them. Other standard Berlinetta equipment included RPO Z54, Interior Dècor/Quiet Sound. This included a glove compartment light, simulated leather accents on the instrument cluster, and additional instrument cluster lighting.

While the Berlinetta came with the base six-cylinder engine, most buyers sprang for one of the optional V-8 engines. The Berlinetta was the most expensive Camaro after the Z28, at $5,395.90. Next in the pecking order was the Rally Sport Coupe, with a base price of $5,072.90. The biggest standout feature of the Rally Sport Coupe from its siblings was the availability of a two-tone paint scheme. Carried over from the preceding year, the option was easily identified. The upper color was separated from the body color by a tri-color decal. With a contrasting upper color, a vinyl roof was not available with the Rally Sport Coupe. Standard on the Rally Sport was the tail spoiler sourced from the Z28, but the performance car's front spoiler was not fitted to the RS. Unlike years prior, none of the four models of Camaro could be mixed with another model. Thus there was no chance for a RS/Z28. But plenty of customers wanted their Camaro in Rally Sport guise, to the tune of 19,101 sold.

Last on the list of Camaros for 1979 was the base Sports Coupe. It listed for $4,676.90, yet the extensive option list could be used to tart it up on a level with the higher-zoot models. Evidently, a lot of people, 111,357 to be exact, felt that way. It was the largest-selling version of Camaro for 1979, and when the last 1979 unit rolled off of the assembly line, a record 282,571 Camaros had been built, a record that stands to this day. Unfortunately for Chevrolet, Ford had released

1979	
MODEL AVAILABILITY	two-door coupe
WHEELBASE	108 inches
LENGTH	197.6 inches
WIDTH	74.5 inches
HEIGHT	49.2 inches
WEIGHT	3,328 lbs
PRICE	$4,676
TRACK	61.3/60.0 inches (front/rear)
WHEELS	14 x 6 inches
TIRES	E78 x 14
CONSTRUCTION	unitized body/frame with bolt-on front subframe
SUSPENSION	long-arm/short-arm with coil springs front/longitudinal leaf springs, live axle rear
STEERING	recirculating ball
BRAKES	11-inch disc front, 9.5 x 2.0 inch drum rear
ENGINE	90- and 115-horsepower, 250-cubic-inch I-6; 125- and 130-horsepower, 305-cubic-inch V-8; 165- and 170- and 175-horsepower, 350-cubic-inch V-8
BORE AND STROKE	3.875 x 3.53 inches (250), 3.736 x 3.48 inches (305), 4.00 x 3.48 inches (350)
COMPRESSION	8.2:1 (250), 8.4:1 (305), 8.2:1 (350)
FUEL DELIVERY	single-barrel (250), single two-barrel (305), single four-barrel (350)
TRANSMISSION	three- and four-speed manual, three-speed automatic Turbo Hydra-Matic
AXLE RATIO	ranging from 2.56:1 to 3.42:1
PRODUCTION	21,913 six-cylinder, 260,658 V-8

a new Mustang for 1979, built on the new Fox platform. Unlike the crappy Pinto-based Mustang II, the 1979 Mustang was a state-of-the-art vehicle, lithe, light, and a head-turner. Sales for the Ford competitor skyrocketed, with the Blue Oval racking up 369,936 sales. The next-generation Camaro was still years away, and the newest iteration of Mustang was a smash in the marketplace. Chevrolet knew that it now had its work cut out for it.

With the use of a full-width tape stripe beneath the tall rear spoiler, Chevrolet ensured that drivers in the wake of the 1979 Camaro Z28 knew what they were following. Access to the gas cap was behind the door in the center of the rear panel.

1980

By the time the 1980 Camaro hit the street, the second generation had been out for a decade. Chevrolet was working had to develop generation three, but it knew that it wouldn't be out until the 1982 model year. Thus the job was to keep the current Camaro interesting and selling while its successor was readied. Needless to say, virtually all of the resources at Chevrolet were directed at the 1982 model. So it was inevitable that significant change on the second-gen cars just wasn't going to be happening.

It was a platform that had been designed in the latter part of the 1960s and, surprisingly, it was still hanging on. But it was getting very long in the tooth. It had not been originally

intended to be loaded with the emission equipment and safety gear that the government was ladling on. When the second-generation debuted, it could be had with a ground-pounding 454. Now the most powerful engine was a 350 rated at 190 horsepower. Yet few automobiles could touch a 1980 Z28 around a skid pad. Chevrolet's engineering staff had honed the Camaro into an excellent canyon carver and long-distance cruiser. That alone is testament to the rightness of the design when it debuted in 1970. For 1980, change was incremental but still significant.

The most important development was the ditching of the venerable straight six-cylinder engine, replaced by a pair of new V-6s. While the old six-pot powerplant had never

garnered headlines, it was a stout but slow performer, able to get good fuel economy while demanding little. It was an ideal engine for a first-time driver, and many youths gazed down the long hood and, in their minds, heard a brawny V-8 while the rest of the world heard an overworked, low-revving engine giving off agricultural sounds. All while delivering agricultural performance. But it got the job done

Yet fuel economy was now the order of the day. In order to improve mileage, Chevrolet ended up using two different 90-degree V-6 engines in the 1980 Camaro. Which one was under the hood depended on where the vehicle was being sold. For six-cylinder Camaros sold in any state but California, an odd-fire Chevrolet-built 3.8-liter, 229-cubic-inch V-6

was used, boasting 115 horsepower. For the denizens of the Golden State, their V-6 Camaro was outfitted with an even-fire Buick-built engine, displacing 3.8 liters, 231 cubic inches, and rated at 110 ponies.

Non-California buyers of the Camaro Sport Coupe, Rally Sport Coupe, and Berlinetta Coupe that demanded a V-8 could step up to RPO L39, displacing 267 cubic inches and generating 120 horsepower. If a few more ponies were needed, then customers would check RPO LG4, with 305 cubic inches in a V-8 configuration. This 155-horse powerplant was available in all 50 states. At the top of the list was the 350-cubic-inch RPO LM1, a 190-horsepower V-8 found in the Z28, except for California. Customers in that state who

While the finned aluminum wheels looked great on the 1979 Camaro Z28, they were a pain in the neck to clean. The standard tire on the Z28 was the Goodyear GR70-15 steel-belted radial. Following normal practice of the day, the Z28 enjoyed 11-inch disc brakes on the front, but 9.5-inch drums on the rear.

A 350-cubic-inch engine developing 190-horsepower was standard in the 49-state Camaro Z28 for 1980. As engine reliability increased, there were fewer reasons to get under the hood, so engine compartment appearance was a diminishing priority for Chevrolet. Yet the Z28 enjoyed a functional hood scoop, as evidenced by the rubber seal on the lip of the air cleaner.

Under heavy cornering, a 1980 Z28 equipped with T-tops suffered a bit of cowl shake when compared with a non-T-top-equipped vehicle, but the vast majority of driving was at speeds less than 10/10ths. This popular option allowed fresh air and sunlight into the cabin, yet the vehicle could be locked up more securely than a traditional convertible.

Body-color bumper caps lent a unified visual look to the 1980 Z28. These were not ideal cars for introverts. Starting retail price for the Z28 was $8,121.32, and with sales of 45,137 units, it continued to be a strong seller.

Left: For 1980, all Z28s, with the exception of California-bound vehicles, were equipped with RPO LM1, a 350-cubic-inch V-8 that generated a whopping 190 horsepower. Z28s that were sold in the Golden State packed RPO LG4 under the hood, its 305-cubic-inches delivering a staggering 155 horses.

Left: Beneath the graceful urethane cover, a government-mandated 5-miles-per-hour bumper worked to reduce body damage under low-speed impacts. The goal was to reduce collision costs, resulting in lower insurance premiums. It didn't quite work out that way; the insurance companies still raised their rates.

Above: With 45,137 Z28s sold in 1980, the second-generation Camaro still appealed to a large number of buyers after more than a decade on the market. Bold tri-color graphics running down the side of the car left no doubt that this was not a grocery-getter Camaro. Chevrolet marketed the car as "The Hugger."

ordered the Z28 found the smaller LG4 engine, rated at 165 horsepower in this usage, between the front wheelwells.

Yet again, manual transmissions were not available in the land of Los Angeles. But in a bit of good news, the Z28 now employed a cold-air cowl induction system. Under normal cruising, rear-facing flaps at the back of the hood were shut; but under heavy throttle, solenoids flipped open the flaps, allowing cool, ambient air to flow in the air cleaner. Performance was not exactly neck-snapping, but a Z28 could at least merge onto a freeway without worry. *Car and Driver* magazine, in their April 1980 issue, printed a review of a Z28 equipped with a four-speed manual transmission and 3.08:1 rear end gears. The headline to the article read "Z28, A medieval warrior on the path to a rocking chair." That assessment might have been a bit harsh, but with a

Above: The tall rear spoiler had originally been released in 1970 as COPO 9796, and it was a popular addition. Starting in 1971, the spoiler was standard on the Z28 and continued virtually unchanged until 1981. One of the downsides to its use was the tendency of the back of the car quickly to be covered in dust.

Right: Unlike the muscle cars of a decade earlier, the 1980 Camaro Z28 shined on twisty roads. From the Z28's inception in 1967, the option had stressed handling prowess as much as brute power, and by the time the 1980 rolled out, the Z28 was one of the finest-handling vehicles in the United States.

Left: The front air dam, extending from one front wheel arch beneath the nose to connect with the other front wheel arch, was new for 1979, and it was effective in managing airflow. Most of the Z28's airflow needs were supplied from under the nose, rather than through the grille. This trend would become industry-wide as the manufacturers strove to reduce frontal drag.

Below: The blacked-out grille, headlight, and parking light treatment made the Red Orange finish (paint code 76) even more striking. The color was applied to 4,374 Camaros in 1980.

quarter-mile result of 16.4 seconds at 86 miles per hour, and a 0–60 miles per hour time of 9.5 seconds, it was clear that the current generation of Camaro was tired and ready for replacement. While it had a top speed of 120 miles per hour, fuel economy was in the 14-mpg range, and bringing the Z28 to a stop from 70 miles per hour took a lengthy 196 feet. In an attempt to slow drivers down, the speedometer was changed from a 130 miles per hour scale to 85 miles per hour.

Minor exterior changes included wire wheel covers on the Berlinetta, new functional front fender vents on the Z28, and the demise of the vinyl roof. In a move to improve

Above: Looking as if it had been dipped in a bucket of blue paint, the monochromatic 1980 Camaro Z28 was an aggressively styled performance car that was vivid and head-turning. The air deflectors in front of the rear wheel arches were "borrowed" from the Pontiac Trans Am.

Right: Most of its 1980 automotive peers got this view of the Z28, especially if the road was full of twists and turns. While horsepower was down in comparison to a decade before, the Z28 of 1980 could easily out-handle its predecessors. A large reason for the F-body's handling prowess was due to beefy stabilizer bars, radial tires, heavy-duty springs, and 1-inch piston shock absorbers with special tuning.

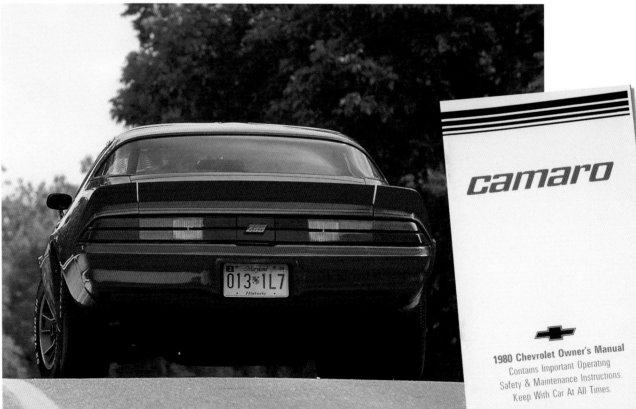

camaro

1980 Chevrolet Owner's Manual
Contains Important Operating
Safety & Maintenance Instructions.
Keep With Car At All Times.

aerodynamics, spats were installed at the front of the rear wheel opening. It was hoped that they would direct air around the rear tires. An added bonus was that they made the Z28 appear even more track ready. Unfortunately, the buying public didn't race to Chevy showrooms. In 1980, the country was gripped by a recession, another Middle East fuel crisis was bubbling, inflation was at a record high, and Americans were watching their wallets. As a result, Camaro sales plunged. Total sales were only 152,005, a significant decline from the preceding year. The Z28 was the second most popular model in the Camaro line, with 45,137 sold. It was clear that there was still a viable market for a performance car, and Chevrolet was close to releasing a new generation of

Below: Horsepower was raised to 180 for 1980, as the LM1 350-cubic-inch V-8 was the strongest in a Camaro since 1974. With a redline of 5000 rpm, the Z28 could cover the quarter-mile in 16.09 seconds at 86 miles per hour. Top speed was 120 miles per hour, more than enough to bury the speedometer.

1980

MODEL AVAILABILITY	two-door coupe
WHEELBASE	108 inches
LENGTH	197.6 inches
WIDTH	74.5 inches
HEIGHT	49.2 inches
WEIGHT	3,328 lbs
PRICE	$5,498
TRACK	61.3/60.0 inches (front/rear)
WHEELS	14 x 6 inches
TIRES	P205-75/14
CONSTRUCTION	unitized body/frame with bolt-on front subframe
SUSPENSION	long-arm/short-arm with coil springs front/longitudinal leaf springs, live axle rear
STEERING	recirculating ball
BRAKES	11-inch disc front, 9.5 x 2.0 inch drum rear
ENGINE	115-horsepower, 229-cubic-inch V-6; 110-horsepower, 231-cubic-inch V-6; 120-horsepower, 267-cubic-inch V-8; 155- and 165-horsepower, 305-cubic-inch V-8; 190-horsepower, 350-cubic-inch V-8
BORE AND STROKE	3.74 x 3.48 inches (229), 3.80 x 3.40 inches (231), 3.50 x 3.48 inches (267), 3.74 x 3.48 inches (305), 4.00 x 3.48 inches (350)
COMPRESSION	8.6:1 (229), 8.0:1 (231), 8.3:1 (267), 8.6:1 (305), 8.2:1 (350)
FUEL DELIVERY	single two-barrel (229, 231, 267), single four-barrel (305, 350)
TRANSMISSION	three- and four-speed manual, three-speed automatic Turbo Hydra-Matic
AXLE RATIO	ranging from 2.56:1 to 3.42:1
PRODUCTION	51,104 six-cylinder, 100,901 V-8

Camaro. But before that happened, the outgoing car had to generate good sales, and Chevy had a few changes in store for 1981.

1981

Right out of the box, the 1981 Camaro lineup was different due to omission. The Rally Sport Coupe was yanked from the menu, leaving the Sport Coupe, the Berlinetta Coupe, and the Z28 Sport Coupe. As usual, the Sport Coupe was the bread and butter version, able to stretch between stripped grocery-getter to near Berlinetta, V-8-equipped grand tourer. The Berlinetta continued to be targeted at the buyer wanting a Camaro with a heavy dose of luxury. And the Z28 was intended to continue to be one of the best-handling cars on American road.

A number of Camaro firsts made their debut on the 1981 models, such as a CCC (Computer Command Control) emission system. This controlled the Rochester Quadrajet carburetor. Another innovation that appeared was a lockup torque converter. It would lock up in second and third gears in the Z28 (third gear only on other models) in an attempt

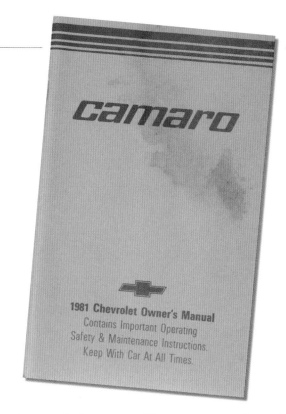

to improve fuel economy. After many years of complaints about the inadequacies of the headlights, halogen bulbs were available as a $36 option.

Under the long hood there were a few changes as well. Finally, California buyers could get a four-speed manual transmission, as long as it was bolted to the 150-horsepower RPO LG4 305-cubic-inch V-8 mounted in the Sport Coupe and Berlinetta Coupe. California Z28 customers who wanted to row their own gears had to settle for the LG4 powerplant, but in the Z28, it was rated at 165 horses, thanks to the cowl induction and a more aggressive exhaust system. More good news for California; the 350-cubic-inch engine was available. But it had to use the Turbo Hydra-Matic automatic transmission, a $61 mandatory option. Power brakes, long a standard feature on Z28s but an option on other models, became standard across the board. In a vehicle that weighed in the neighborhood of 3,300 pounds for the Sport Coupe and Berlinetta, and 3,600 pounds for the Z28, power brakes were a very good thing.

Sales in the second generation's last year were better than expected, with a total of 126,139 units being shown the door. After 12 years, Chevy had definitely gotten its investment out of this run. Yet its competitors had passed it by, and it was time to greet a new generation. When the 1982 Camaro hit the showrooms, it was evident that Chevrolet was serious about the pony car market, and the Camaro was suddenly a contender. But that's for another chapter. . . .

1981	
MODEL AVAILABILITY	two-door coupe
WHEELBASE	108 inches
LENGTH	197.6 inches
WIDTH	74.5 inches
HEIGHT	49.2 inches
WEIGHT	3,328 lbs
PRICE	$6,581
TRACK	61.3/60.0 inches (front/rear)
WHEELS	14 x 6 inches
TIRES	P205 75/14
CONSTRUCTION	unitized body/frame with bolt-on front subframe
SUSPENSION	long-arm/short-arm with coil springs front/longitudinal leaf springs, live axle rear
STEERING	recirculating ball
BRAKES	11-inch disc front, 9.5 x 2.0 inch drum rear
ENGINE	110-horsepower, 229-cubic-inch V-6; 110-horsepower, 231-cubic-inch V-6; 115-horsepower, 267-cubic-inch V-8; 150- & 165-horsepower, 305-cubic-inch V-8; 175-horsepower, 350-cubic-inch V-8
BORE AND STROKE	3.74 x 3.48 inches (229), 3.80 x 3.40 inches (231), 3.50 x 3.48 inches (267), 3.74 x 3.48 (305), 4.00 x 3.48 (350)
COMPRESSION	8.6:1 (229), 8.0:1 (231), 8.3:1 (267), 8.6:1 (305), 8.2:1 (350)
FUEL DELIVERY	single two-barrel (229, 231, 267), single four-barrel (305, 350)
TRANSMISSION	three- and four-speed manual, three-speed automatic Turbo Hydra-Matic
AXLE RATIO	ranging from 2.56:1 to 3.42:1
PRODUCTION	52,004 six-cylinder, 74,135 V-8

Right: The speedometer in the 1981 Camaro read only to 85 miles per hour, a pathetic attempt by the government to modify the driving public's behavior. Their rationale was that if the speedometer didn't display triple digits, a driver would be less likely to exceed the speed limit, thus saving fuel.

Above: The 1981 Camaro Sport Coupe was the last of the second-generation Camaros. It was still a very affordable car with a retail price just under $6,000, and it was the most popular Camaro model sold that year, with more than 62,000 going to good homes. Yet, the public was awaiting the all-new Camaro for 1983.

GENERATION THREE

1982–1992 Less Is More

Chevrolet tipped over the apple cart in 1982 when it debuted the third-generation Camaro. For decades, Detroit auto manufacturers had lived by the rule that each new generation of any vehicle had to be longer, wider, and heavier than its predecessor. But that mold was shattered when the 1982 hit the street. It rode on a seven-inch-shorter wheelbase than the 1981 version and tipped the scales some 500 pounds lighter. Its weight was within 70 pounds of the original 1967 Camaro, a remarkable achievement.

It almost didn't happen. Development work on the third generation started in 1975, and, for a time, there was a very real possibility that the Camaro would become a front-drive automobile. In the mid-1970s, front-wheel drive was making dramatic inroads into the public's garages, and General Motors had a vehicle, the X-car, that they felt could do battle with the hordes of imports. General Motors had purchased a Volkswagen Scirocco to evaluate a sporting front-drive vehicle. There were proponents within Chevrolet who felt that

The standard engine in the 1982 Camaro Indianapolis 500 pace car replica was the carbureted LG4 5.0-liter V-8, rated at 145 horsepower. The only optional engine was the fuel-injected LU5 5.0-liter V-8, cranking out 165 horsepower. The actual track cars had 250 horsepower aluminum-block engines.

Early design exercises for the third-generation Camaro looked almost like mid-engine sports cars. *General Motors 2012*

the next-generation Camaro should follow the Scirocco's lead. Chevrolet's version of the front-drive X-car was the Citation, a distinctly un-sporty vehicle. But the front-drive camp felt that many of the X-car components could be used in a new Camaro.

Fortunately, cooler heads prevailed, particularly when word came down from on high that the third-generation Camaro had to be able to use a V-8. In the 1970s world of pony cars, it was essential that a V-8 be available. Even the pathetic Mustang II offered a V-8. Two design teams worked on the newest Camaro, as recounted in Anthony Young's excellent book *Camaro*; "Bill Porter's Advanced No. 1 studio started its work on the F-body layout and configuration, while Jerry Palmer's Chevrolet No. 3 studio focused on establishing the new Camaro's identity in light of its inevitably small exterior envelope and proposed front-wheel drive configuration." The

vehicle was designed to fit either a front- or rear-drive system, but the X-car bits that were intended to make a front-drive Camaro a reality just weren't strong enough to handle a V-8. Too bad, so sad.

Camaro chief engineer Tom Zimmer put it well in Gary Witzenburg's book *Camaro: An American Icon*. "In spite of the obvious trend to front-wheel drive, when we aligned our priorities and looked as the customer and the market we were trying to serve, the facilities, the packaging, the mass, the fuel economy, and all those things, and looked at what kind of physical arrangement would produce the kind of car we wanted, we said we wanted to make it rear-wheel drive." Then, as now, the fundamental essence of Camaro was front engine, rear drive.

In August 1977, GM Design Vice President Bill Mitchell retired and was replaced by Irv Rybicki. Rybicki ordered a review of all designs in the pipeline, and when the third-generation Camaro came under his gaze, he requested a new direction. In Bill Porter's studio, an assistant named Roger Hughet created a sketch of what he wanted a Camaro to look like. Gary Witzenburg recalls Rybicki's response: "We took one look at it and . . . said, 'Do it in scale.' Then we tried it full-size, and it was a success from the very beginning. We didn't think we could get there because this car is considerably shorter and narrower, has better seating in back, a hatch, and fold-down rear seats. And we've got one design feature on it that no one else has ever done. We take the backlight glass and form it into a compound curve in two directions, and then it flows right into the sheetmetal surfaces. There is a deck, but you never quite see the break." One of the features of the

second-generation Camaro that was carried over was the large, single piece of side glass. Irv Rybicki, General Motors' vice president of design, liked the large Day Light Opening (DLO), feeling that it was a simple way to impart a sporty greenhouse. Unfortunately, the design teams had not planned for that styling device, so significant work had to be redone. By the time that issue, as well as myriad other details, had been hammered out, the planned release date of the third-generation Camaro, 1980, had to be pushed back to 1982. This gave Chevrolet a bit more time to hone the newest Camaro. When it did hit the street, it was found that it was time well spent. The Camaro leapfrogged to the top of the heap.

It was named "Car of the Year" by *Motor Trend* magazine, evidence that Americans loved their aggressive-looking,

In 1982, a Camaro once again paced the Indianapolis 500, this time to commemorate the introduction of the newest Camaro. The Z28 that was used on the track needed no modifications to handle its pace car chores. The NACA ducts on the hood were nonfunctional, but they were a significant styling feature. The complex and huge rear window/hatch had a habit of shattering. Chevrolet had to replace a lot of windows until the supplier was able to work the bugs out and make a dimensionally stable window.

Chevrolet sold 6,360 Indianapolis 500 pace cars replicas, RPO Z50 ($10,600). All of the pace car replicas were built in Chevrolet's Van Nuys, California, plant between mid-March and late April. The pace car would cover the quarter-mile in 16.4 seconds at 86 miles per hour.

The huge overhang in front of the wheels made parking a challenge, and the approach angle wasn't the best. But the long, tapering bumper car helped provide the 1982 Z28 Indianapolis pace car with superb aerodynamics. The strobe stripe wrapping fully around the car was available only on the pace car edition.

sporty cars. The editors of the magazine wrote, "It's a bold move, committing to an all-new performance machine when everyone else is thinking economy. The new Camaro boasts what is likely the most carefully developed 'handling' chassis ever issued by Detroit, as well as daring sheet metal. The Z28 is the hard-core Camaro, offering proper suspension, 4-wheel discs, big tires and quick steering. Recaro-like support comes from the excellent new Conteur driver's seat. If you're making up your personal shopping list of great road cars and you don't have a Z28 or Trans Am on it, you need a new list. The Camaro is simply the best American road car ever built. When one car can meld supreme roadworthiness, dramatic styling and contemporary engineering in an exciting package that sells for $10,000, it deserves unique recognition."

Being put on a diet was part of the plan to catch a competitor. But it wasn't the Mustang. No, the bogey was the Pontiac Trans Am, particularly when equipped with the WS-6 handling package. The new Camaro, any model, could be equipped with optional four-wheel disc brakes for the first time since 1969. This went a long way in making the new Camaro a world-class handling machine. Carrying a significantly lighter body did wonders for the Camaro's road manners too, as well as delivering the all-important excellent fuel economy. The lines of the car were crisper, leaner, and with the huge compound curve rear window, which was hinged to lift up and provide access to the large storage area, the 1982 Camaro cut a rakish profile. Early examples encountered a problem, as the rear window, the largest piece of glass in automotive use up to that time, had a disconcerting habit of exploding when lowered too briskly.

But further development by the glass supplier soon solved the problem, and drivers found that blind spots were virtually non-existent.

The structure of the new Camaro departed from its predecessor. That car used a combination unibody and subframes. The 1982 Camaro boasted a complete unit body, which gave the platform increased rigidity as well as light weight. This allowed the engineers to fit a modern MacPherson strut front suspension, giving the car tauter, more responsive reflexes. A solid rear axle was retained, but the leaf springs were tossed into the dumpster, replaced with lower control arms made of stamped sheet metal, lateral track bars, fore-aft torque arms, and coil springs. This rugged configuration was controllable, as well as able to handle any power increases the future might hold.

Tom Zimmer was the Camaro chief engineer, and he and his team started work on the engineering direction of the next-gen F-body in 1976. Gary Witzenburg spoke with Zimmer. "In gross terms," he recalled, "the objectives were to significantly reduce the weight of the car; try to keep the seating for the front passengers essentially unchanged; improve, or at least maintain, the rear seating; significantly improve the luggage capacity; maintain a specialty-car character, high-styled with a sporty image; and improve the fuel economy. We felt we knew going in what the mission of this car ought to be. It ought to be an extremely good-handling car . . . [W]e used benchmarks like the Lotus Esprit and the Porsche 924."

The suspension wasn't the only change to the Camaro. The engine compartment was the scene of progress. For the first time in the Camaro's history, an inline four-cylinder powerplant was fitted. Called the "Iron Duke," it was "borrowed" from Pontiac and displaced 151 cubic inches. Its 90 horsepower wasn't going to ignite much passion, but its excellent fuel economy kept the government off of Chevy's back. It used a throttle-body fuel-injection unit to improve drivability and better mpg. The Iron Duke was the base engine in 1982 and proved agreeable with 21,802 buyers. For customers who actually wanted to get out of their own way, a 173-cubic-inch V-6, pumping out a fearsome 102 horsepower, could be slipped under the hood. This engine was the standard mill in the luxury-heavy Berlinetta model.

Of course, the Camaro and performance went together like peas and carrots, and for $10,099 base price, you could have your name on the title of a Z28. The 350-cubic-inch V-8 that had been propelling the 1981 Z28 was no longer on the menu; for 1982, a pair of 5.0-liter, 305-cubic-inch V-8s were on the roster. When a customer ordered a standard Z28, the engine was topped by a four-barrel carburetor, and in that configuration it belted out 145 horsepower. Yet another new-for-Camaro feature pointed toward the future, fuel injection. Called Throttle Body Injection (TBI), RPO LU4 used the name Cross-Fire Injection, was rated at 165 horsepower, and was only available on the Z28. This option cost $450 and was rather popular, as 24,673 were installed in the Z28. It was only available with an automatic transmission initially, but later in the model year, a manual transmission could be bolted to the back of the block.

Not for the first time, the Camaro was tapped to act as Pace Car at the annual Indianapolis 500-Mile Race. This was the third time the F-body had led the pack around the 2½-mile course, and two actual Pace Cars were built at the Van Nuys, California, plant, equipped with all-aluminum 350-cubic-inch engines. The customer versions of the Pace

Tasteful tri-color taillights wrapped around the side of the 1982 Z28 Indianapolis 500 pace car. It had taken a few years, but Chevrolet's designers had mastered the art of making a federally mandated bumper look attractive. The small rear spoiler didn't really generate any useable downforce, but it fit the look of the car very well.

1982

MODEL AVAILABILITY	two-door coupe
WHEELBASE	101.1 inches
LENGTH	187.8 inches
WIDTH	72.0 inches
HEIGHT	49.8 inches
WEIGHT	2,875 lbs
PRICE	$8,029
TRACK	60.7/61.5 inches (front/rear)
WHEELS	14 x 6 inches
TIRES	P195/75R-14
CONSTRUCTION	unit body
SUSPENSION	modified MacPherson strut, coil springs front/torque arm, struts, coil springs rear
STEERING	recirculating ball
BRAKES	10.5-inch disc front/9.5 x 2.0-inch drum rear
ENGINE	90-horsepower, 2.5-liter I-4; 102-horsepower, 2.8-liter V-6; 145- and 165-horsepower, 5.0-liter V-8
BORE AND STROKE	4.00 x 3.00 inches (2.5-liter), 3.50 x 2.99 inches (2.8-liter), 3.74 x 3.48 (5.0-liter)
COMPRESSION	8.2:1 (2.5-liter), 8.5:1 (2.8-liter), 8.6:1 (145-horsepower 5.0-liter), 9.5:1 (165-horsepower 5.0-liter)
FUEL DELIVERY	single two-barrel (2.5-liter, 2.8-lier), TBI (2.5-liter), single four-barrel (145-horsepower 5.0-liter), CFI (165-horsepower 5.0-liter)
TRANSMISSION	four-speed manual, three-speed automatic
AXLE RATIO	ranging from 2.73:1 to 3,23:1
PRODUCTION	21,802 four-cylinder, 69,777 six-cylinder, 98,168 V-8

A plastic cap in the center of the wheel hid the lug nuts, giving the wheel a finished appearance, important when the actual 1982 Z28 Indianapolis 500 pace car was seen by millions of people. Shod with Goodyear Eagle GT tires, the subtle red stripe on the rim of the wheel was unique to the pace car package.

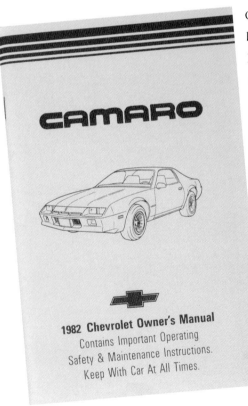

CAMARO

1982 Chevrolet Owner's Manual
Contains Important Operating Safety & Maintenance Instructions. Keep With Car At All Times.

Car were also assembled at the Van Nuys plant, but engine choices were the same as regular Z28s. Every Pace Car replica had identical paint to the actual pace cars, silver over blue. Evidently, quite a few buyers had visions of rolling across the yard of bricks, as 6,360 of the $10,999 Pace Car replicas were sold.

The exterior of the newest Camaro wasn't the only area that was dramatically updated. The interior was now fresh, cleanly styled, and much more comfortable. In an interior styling studio under John Shettler, an unusual number of sources lent inspiration to the designers. Business jets and the Concorde were studied for the methods to control glare and reflections from a sharply angled windshield. The new Camaro's windshield was canted back a staggering 62 degrees, and reflections were a concern. The aircraft used instrument panels that were heavily hooded to reduce glare, and this is the path Shettler's crew took. The instrument

panel was far easier to read, and a unique twin-needle speedometer showed both mph and kph simultaneously by using opposite ends of the same indicator needle. The front seats now came with a reclining feature standard, and an optional Z28-only Lear-Siegler-built front driver seat, called LS Conteur, was available. This single seat had a full range of adjustable features, including thigh support, lumbar, and backrest bolster. Depending on the option package, it could cost up to $611.

Overall performance rivaled some of the world's best sports cars, with handling that put the 1982 Z28 at the top of the list for American vehicles. In a straight line, acceleration was a shadow of former days, but then every carmaker was struggling to extract low ETs. *Motor Trend* magazine even they admitted that a touch more briskness would be appreciated. It took *MT*'s lead-footed editors 17.13 seconds to cover the drag strip. The staff at *Car and Driver* recorded a 0–60 time of 9.7 seconds, at a time when the Mustang GT required 8.0. The crew at Chevrolet knew that they still had a lot of work to do.

With 189,747 built, the 1982 Camaro got the third generation off to a good start. Once again, the Camaro had beaten the Mustang in sales. The Blue Oval recorded 'Stang sales in model year 1982 of 130,418 units, far behind the Bowtie. So Chevrolet followed the axiom "don't mess with a

The third-generation Camaro didn't utilize any radical engineering, but it was the first clean-sheet redesign of the Camaro in the car's 25-year history. *General Motors 2012*

good thing" when it came to the 1983 Camaro. Inevitability, some color choices were modified, but major changes weren't in the cards. New for 1983 was a five-speed manual transmission, standard now on the Berlinetta and Z28. In April 1983, a new engine appeared, the RPO L69 5.0-liter HO. Priced identically as the 5.0 TPI at $450, the HO used a single Rochester four-barrel carburetor atop an aluminum intake manifold, a special camshaft, dual intake air cleaner with ducting in the grille feeding the engine, and was rated at 190 horsepower. Only one transmission was available with the L69, the Borg-Warner T-5 five-speed manual. The addition of horsepower did wonders for the acceleration of the Z28; 0–60 times now came in a tick over 7 seconds, while it ran the quarter-mile in 15 seconds flat. This was the most powerful engine in available that year, and even though the powerplant was offered late in the model year, 3,223 were sold. For performance enthusiasts, it was a clear sign that Chevrolet was listening to them.

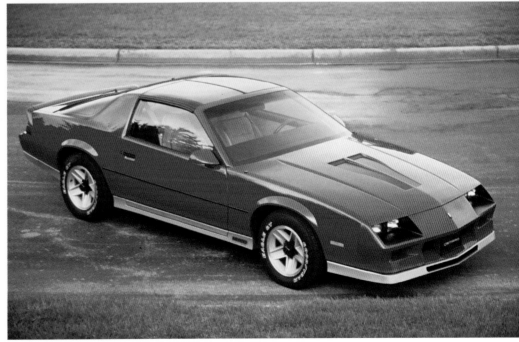

Any performance gain for the 1982 Z28 was the result of reduced weight rather than increased power, but the Camaro would soon receive a much-needed injection of horsepower. *General Motors 2012*

Above: The Berlinetta Coupe was targeted at buyers who were more interested in sporty looks than raw performance. *General Motors 2012*

Right: The Iron Duke—the 151-cubic-inch, four-cylinder that was the base engine in the 1982 Camaro—was an antiquated design when Pontiac introduced it in 1977. It hadn't aged well in the intervening years. *General Motors 2012*

The bucket seats in the 1983 Z28 took their design cues from the Recaro seats used in high-end European sports cars of the period. *General Motors 2012*

1983

Sales in the third generation's sophomore year softened a bit, with total numbers coming in at 154,381 units. Ford would have loved to have had similar sales numbers for the Mustang, with 120,873 for Dearborn. But a closer look shows that the public was really responding to the increased emphasis on power. Z28 sales were 62,650, just a whisper below the bread-and-butter Sports Coupe's 63,806. Chevrolet loved seeing so many of the high-margin Z28s rolling out of showrooms; the Sport Coupe model was a much lower profit vehicle. Tragically, the 8-track stereo system was no longer available in 1983. Cassettes were the new big thing in listening convenience.

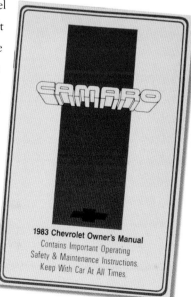

1983 Chevrolet Owner's Manual
Contains Important Operating Safety & Maintenance Instructions. Keep With Car At All Times.

1983	
MODEL AVAILABILITY	two-door coupe
WHEELBASE	101.1 inches
LENGTH	187.8 inches
WIDTH	72.0 inches
HEIGHT	49.8 inches
WEIGHT	2,878 lbs
PRICE	$8,036
TRACK	60.7/61.5 inches (front/rear)
WHEELS	14 x 6 inches
TIRES	P195/75R-14
CONSTRUCTION	unit body
SUSPENSION	modified MacPherson strut, coil springs front/torque arm, struts, coil springs rear
STEERING	recirculating ball
BRAKES	10.5-inch disc front/9.5 x 2.0-inch drum rear
ENGINE	92-horsepower, 2.5-liter I-4; 107-horsepower, 2.8-liter V-6; 150-, 175-, 190-horsepower, 5.0-liter V-8
BORE AND STROKE	4.00 x 3.00 inches (2.5-liter), 3.50 x 2.99 inches (2.8-liter), 3.74 x 3.48 inches (5.0-liter)
COMPRESSION	8.2:1 (2.5-liter), 8.5:1 (2.8-liter), 8.6:1 (150-horsepower, 5.0-liter), 9.5:1 (175- & 190-horsepower, 5.0-liter)
FUEL DELIVERY	single two-barrel (2.5-liter LQ8, 2.8-liter), TBI (2.5-liter LQ9), single four-barrel (150- & 190-horsepower, 5.0-liter), CFI (175-horsepower, 5.0-liter)
TRANSMISSION	four- and five-speed manual, three- and four-speed automatic
AXLE RATIO	ranging from 2.73:1 to 3.73:1
PRODUCTION	9,926 four-cylinder, 54,332 six-cylinder, 90,123 V-8

While Chevrolet hadn't built a convertible Camaro since 1969, the removable T-top roof provided some semblance of open-air motoring. *General Motors 2012*

1984

When the 1984 Camaro rolled out, the biggest news was what was missing. That would be the 5.0-liter TBI engine. Word had gotten out among enthusiasts that for the same money as the advanced fuel injection system, a simpler, more powerful mill could provide mucho thrills. For Chevy, pulling the TBI engine was a no-brainer. General Motors' engineering staff was working on the next generation of fuel injection, so a couple more years of carbureted engines wasn't a problem. However, the cowl-induction system was pulled from production, as water entering the induction system was a problem.

The rest of the engine lineup was carried over from the preceding year, including the Iron Duke four-cylinder engine. Only one model of Camaro, the Sports Coupe, would be fitted with a four-speed manual transmission, and that box was only bolted to the four-banger. This would be the final year

for a four-speed manual transmission. Only 10,687 Camaros were equipped with the 151-cubic-inch straight four, now rated at 92 horsepower. While sales of the Camaro with the smallest engine had exceeded 20,000 units in 1982, by 1984 it was clear that the public was willing to spend a little more to get a vehicle that had some fast behind the flash. The 173-cubic-inch V-6, generating a lofty 107 horsepower, was standard in the Berlinetta and available in the Sports Coupe. This proved to be very popular, as a total of 98,471 Camaros were so equipped in 1984. Performance was a malleable term, and the Berlinetta was evidence that a vehicle didn't have to possess tire-melting ability to succeed in the market. With a 16.99-second time in the quarter-mile at 81.1 miles per hour, and a 0–60 time of 9.25, the tires were very safe.

With all of the luxury features standard on the Berlinetta, it needed all of the grunt it could get to roll down the road. Ever since the third generation had debuted, Chevrolet had

Above: Slippery bodywork was only part of the 1984 Z28's success. Equipped with a standard 5.0-liter engine rated at 150 horsepower, it could be fitted with an optional RPO L-69 HO 190-horse mill that enjoyed a compression bump from 8.6:1 to 9.5:1, and was topped with a 650-cfm carburetor.

Left: It seems everyone wanted a Z28 in 1984, what with 52,457 being built with the RPO L69 5.0-liter V-8. This optional engine was only available on the Z28 and cost $530. It was rated at 190 horsepower and breathed through a four-barrel Rochester Quadrajet carburetor. A popular engine, 52,457 were installed in Z28s.

The 5.0-liter V-8 in the 1984 Z28 was rated at 150 horsepower, and it would propel the pony car down a drag strip in 16.4 seconds at 86 miles per hour. Engine compartments were becoming somber looking, as the increased reliability of powerplants reduced the need to get under the hood.

been pushing the Z28. Now that the Z28 was well established as the premier pony car, Chevy shifted its attention to the Berlinetta, equipping it with a unique instrument panel inspired by the Corvette. Light-emitting diodes had entered the commercial market, and this "space-age" technology was used in everything from watches and toasters to automobiles. Chevrolet used the LEDs to create "cutting-edge" displays in their sports cars, and the public responded with a love/

hate relationship. This panel featured digital readouts and an electronic bar tachometer; when dark, it looked like the Ginza at night. A pair of radios, unique to the Berlinetta, carried the digital theme as well. Two adjustable pods, accessible by fingertip, were near the steering wheel, allowing the driver to access radio functions, lights, turn signals, and wipers without removing a hand from the rim of the wheel.

Yet with all of the improvements to the Berlinetta, the Z28 continued to reign supreme. Sales for the performance Camaro set a record that has never been topped: 100,899 units. Total Camaro sales for model year 1984 were a very impressive 261,591. That would be the third-best sales year in Camaro history and the best sales performance for the entire third generation. To Chevrolet, this was a very clear indicator that a concerted push for more power would result in more sales.

Fortunately, the engineers were making serious advances toward integrating computers into a growing list of jobs on a vehicle. As fuel-injection systems improved, more sensors fed information to the computers handling the fuel/air mixture, giving the engineers increased control over the combustion process over a wide range of conditions. This resulted in the enthusiast's trifecta: better performance, improved fuel economy, and reduced emissions. As the decade rolled on, remarkable gains would be seen in all three areas, creating an ever-more exciting Camaro.

1985

That effort would become evident in the 1985 model year. The Iron Duke engine, in the first three months of sales, only saw duty in 3,318 Sport Coupes. Chevrolet responded by pulling the plug on the cast-iron boat anchor. The standard Berlinetta engine, the 173-cubic-inch V-6, got a significant upgrade with the installation of Multi-Port Fuel Injection (MFI). Besides improving the drivability and fuel economy of the engine, power was increased from 107 horsepower to 135. Total V-6 sales were robust, with 78,315 units sold. But changes in the world of the Z28 were profound, and the visibility of the brand would skyrocket.

Speaking of visibility, the Camaro wore a new nose for 1985, with a new spoiler and valance on the Z28, and a new sub-model of the Z28, called the IROC, which was equipped with deep side "ground effect" body cladding and plenty of graphics. Chevrolet, like every domestic manufacturer, knew

1984	
MODEL AVAILABILITY	two-door coupe
WHEELBASE	101.1 inches
LENGTH	187.8 inches
WIDTH	72.0 inches
HEIGHT	49.8 inches
WEIGHT	2,907 lbs
PRICE	$7,995
TRACK	60.7/61.5 inches (front/rear)
WHEELS	14 x 6 inches
TIRES	P195/75R-14
CONSTRUCTION	unit body
SUSPENSION	modified MacPherson strut, coil springs front/torque arm, struts, coil springs rear
STEERING	recirculating ball
BRAKES	10.5-inch disc front/9.5 x 2.0-inch drum rear
ENGINE	92-horsepower, 2.5-liter I-4; 107-horsepower, 2.8-liter V-6; 150- and 190-horsepower, 5.0-liter V-8
BORE AND STROKE	4.00 x 3.00 inches (2.5-liter), 3.50 x 2.99 inches (2.8-liter), 3.74 x 3.48 inches (305-liter)
COMPRESSION	9.0:1 (2.5-liter), 8.5:1 (2.8-liter), 8.6:1 (150-horsepower, 5.0-liter), 9.5:1 (190-horsepower, 5.0-liter)
FUEL DELIVERY	TBI (2.5-liter), single two-barrel (2.8-liter), single four-barrel (5.0-liter)
TRANSMISSION	five-speed manual, four- and five-speed automatic
AXLE RATIO	ranging from 3.08:1 to 3.73:1
PRODUCTION	10,687 four-cylinder, 98,471 V-6, 152,433 V-8

Left: 1984 was a banner year for Z28 sales, with 100,899 snapped up. Never before or since has the Z28 racked up such numbers. And it wasn't a fluke: the 1984 Camaro hit the mark spot on. The Z28 topped numerous auto magazines lists of best sporty car. With its sleek nose, the coefficient of drag was on 0.339, making it one of the slipperier cars sold in America.

Below: The Camaro was massively popular for 1984, with a total of 261,591 built. Of that sum, 100,899 were Z28s, with a base price of $10,620. Unlike prior years that packed on the pounds, the third-gen Camaro weighed 70 pounds *less* than a 1967 Camaro!

In 1985, the $659 B4Z IROC sports equipment package could be installed only on Z28s. Right off the showroom, the Z28 IROC was a serious performance car, capable of delivering 0.92g of lateral force, while sprinting to 60 miles per hour in the 7-second range.

that the public's interest in any automobile was a fickle thing. To the marketing staff, they approached the task of selling automobiles using a time-honored maxim; newer is better. "Planned obsolescence" had been part of the General Motors approach for decades, and here was yet another example that the practice was still alive and well. As the Camaro's third-generation started getting a few years notched on its belt, the marketing department felt that by having the Design department create a new "look" for the newest Camaro, people would be inclined to trade in their two- and three-year-old models to have the "latest and greatest." This game of bait and switch had been going on since the 1920s, and in 1985, it was business as usual.

For 1985, Chevrolet signed on to be the primary sponsor of a race series called International Race of Champions, IROC for short. The premise of the series was to put the world's best race car drivers, from a broad selection of motorsports arenas, in identically prepared vehicles. Each car was brightly colored, the better to show up on television. For a number of years, Porsche supplied 911s to the series; but starting in 1975, the Bowtie was the supplier of the race vehicles. The Camaro was tapped as the basis for the race car, and the series ended up being a good spectacle. Chevrolet had gotten a positive response from the Indy Pace Car program in 1982, so it was with little hesitation that the street Z28 became the defacto race car. But Chevy used a page from the Pace Car replica playbook and released a street version of the IROC package as an option for the Z28.

Priced at $659, option B4Z was one of the best performance bargains ever offered. For that sum, a Z28 received a lot of upgrades, including special Delco-Bilstein shock

continued on page 169

Above: Vibrant seat covers carried over the red and black interior theme of the 1985 IROC-Z. Engineers fitted the catalytic converter beneath the raised portion of the floor in front of the passenger seat, compromising legroom. While some viewed the interior as austere, others called it professional.

Left: Per the fashion of the time, the interior of the 1985 IROC-Z was all crisp lines and somber tones. Yet this purposefulness paid off in allowing a driver to concentrate on the art of driving, rather than searching the instrument panel for information amid a heavily stylized design.

The Z28 moniker, which started with the 1967 model Camaro, was not a tribute to Zora Arkus-Duntov, as some have forwarded. It was simply the internal code denoting a specific performance option package. Chevrolet's marketing staff felt that the Z/28 code sounded strong, so that became the name of the sportiest Camaro.

Right: Design of a vehicle could be improved significantly by designing the front as smooth as possible. That resulted in nearly flush parking lights and a minimum of open grille. Cooling air would be drawn from beneath the long nose.

Opposite: Chevrolet's engineers raised the Z28's performance and handling envelope in 1985 with the introduction of the IROC-Z. With its sticky Goodyear Gatorbacks and Delco front struts incorporating special valving and Bilstein shock absorbers, it was one of the finest-handling automobiles on American roads in the mid-1980s.

Right: Not a bit of chrome in sight, the interior of the 1985 Z28 IROC was a workplace for generating velocity. The beefy shifter fell readily to hand, and the ventilation and radio controls were logically placed. The primary gauges were large and easy to read, and the secondary instruments were centrally located.

Below: Chevrolet's engineers mounted the third-generation Z28's engine as far back as possible in an effort to improve the vehicle's performance by moving the vehicle's center of gravity as close to the center of the car as possible. With the rear portion of the engine underneath the bottom of the windshield, servicing the powerplant could be a challenge.

Rolling stock on the 1985 Z28 IROC's were 16 x 8 cast-aluminum wheels, surrounded by P245/50VR16 Goodyear Eagle Gatorback radial tires. Retaining a five-spoke pattern of years before, they were stylistically freshened up with crisper lines, and the wheels were color-matched to the body color.

absorbers, specific front springs with increased rates, struts and jounce dampers, bespoke rear stabilizer bar and springs, a front frame rail reinforcement called the "Wonder Bar," higher effort steering due to special valving in the steering box, revised camber settings, 5 degrees of caster, an increase from the 3.5 degrees on the standard Z28, Goodyear Eagle P245/50VR high-performance tires mounted on aluminum 16 x 8 wheels, fog lamps, and assorted graphics. The IROC-Z28 sat ½-inch lower than a standard Z28. On a skid pad, the IROC could generate a lateral force of 0.92g, and it would sprint to 60 miles per hour in 6.87 seconds. The quarter-mile was dispensed with in 15.5 seconds, at 89.1 miles per hour. A slalom course returned a 63.3-miles per hour run, very close to Corvette territory.

Two engine choices were available for the IROC-Z: the carbureted RPO L69 V-8 and the new RPO LB9 V-8 with tuned port injection. Buyers springing for the L69 engine had to accept the manual transmission, and vehicles with TPI had to have an automatic tranny, as the manual box couldn't handle the torque. The IROC-Z option was surprisingly popular, considering its price. A total of 21,177 were built, as enthusiasts quickly appreciated just how good a performance car it was.

Above: In the 1980s, American automobile manufacturers started listing engine displacement using the metric system, turning the 350-cubic-inch V-8 engine found in the 1987 Z28 IROC into a 5.7-liter powerplant.

Below: The Z28 benefited from a nose job in 1985, a change made more obvious on the IROC-Z, thanks to the car's handsome monochromatic paint job. With its modified suspension—special rear springs and a stronger rear stabilizer bar—the ride height was lowered by a half inch, improving what was already the best-handling car in America.

1985

MODEL AVAILABILITY	two-door coupe
WHEELBASE	101.1 inches
LENGTH	187.8 inches
WIDTH	72.0 inches
HEIGHT	49.8 inches
WEIGHT	2,907 lbs
PRICE	$8,363
TRACK	60.7/61.5 inches (front/rear)
WHEELS	14 x 6 inches
TIRES	P195/75R-14
CONSTRUCTION	unit body
SUSPENSION	modified MacPherson strut, coil springs front/torque arm, struts, coil springs rear
STEERING	recirculating ball
BRAKES	10.5-inch disc front/9.5 x 2.0-inch drum rear
ENGINE	88-horsepower, 2.5-liter I-4; 135-horsepower, 2.8-liter V-6; 155-, 190-, 215-horsepower, 5.0-liter V-8
BORE AND STROKE	4.00 x 3.00 inches (2.5-liter), 3.50 x 2.99 inches (2.8-liter) 3.74 x 3.48 inches (5.0-liter)
COMPRESSION	9.0:1 (2.5-liter), 8.9:1 (2.8-liter), 8.6:1 (155-horsepower, 5.0-liter), 9.5:1 (190-, 215-horsepower, 5.0-liter)
FUEL DELIVERY	TBI (2.5-liter), MFI (2.8-liter), single four-barrel (155-, 190-horsepower, 5.0-liter), TPI (215-horsepower, 5.0-liter)
TRANSMISSION	four- and five-speed manual, four-speed automatic
AXLE RATIO	ranging from 3.08:1 to 3.73:1
PRODUCTION	3,318 four-cylinder, 78,315 V-6, 98,385 V-8

First year for the tuned port injection, the 1987 Z28 IROC was not an inexpensive automobile, as the coupe went for $13,488, and a breathtaking $17,917 for a convertible. The 90-degree 5.7-liter V-8 generated 225 horsepower at 4,400 revs, while torque came in at 330 lb-ft at 2,800 rpm.

By the time the 1985 Camaro wrapped production, 180,018 had been built. Yet Chevrolet wasn't finished burnishing their pony car. They felt that the Camaro had acquitted itself well in the pony car battles so far, and it was hoped that better days were ahead. To make that a reality, Chevrolet took the axe to some parts of the Camaro lineup, pruning non-performing segments, to concentrate on the types of vehicles the public was clamoring for. The IROC-Z28 was in fact threatening the Corvette for overall performance, and with the +2 rear seats and strangely shaped but large storage area, the Camaro enjoyed more real-world livability than the 'Vette. Factor in the lower purchase price of the F-body, as well as the reduced insurance premiums, and it was clear to see why the Camaro was selling in significant numbers.

1986

The first thing most people noticed about the 1986 Camaro was the goiter-like growth at the top of the rear window. Federal regulations required that all passenger vehicles sold in the United States, starting in 1986, had to display a center high-mount stop lamp (CHMSL). But it didn't take long for interested parties to notice other changes to the Camaro. Gone was the anemic Iron Duke four-cylinder engine. It wasn't missed. Also replaced was the standard four-speed manual transmission, its place taken up by a five-speed manual box. Shortly after production began, Chevrolet ceased building the Berlinetta. Sales of the luxury-oriented vehicle had been soft and were getting softer. It was very clear that the return of serious performance to the Camaro stable was pushing buyers' buttons, not a luxo-Camaro. Only 4,479 Berlinettas were constructed before that party was declared over. Another loss for 1986 was the discontinuation of RPO L69 after only 74 were built. The remaining high-performance engine, the TPI 305-cubic-inch V-8, saw its output reduced from 215 horsepower to 190 due to a camshaft change.

On the plus side, a self-dimming interior mirror was now available, as were rear window slats. This was a year of catching one's breath, as Chevrolet prepared the next major performance jump. Yet the public loved what they saw for 1986, buying a total of 192,219 Camaros, including 88,132 Z28s and IROC-Zs. That turned out to be the second-highest selling year ever for the Z28.

Left: In 1986, buyers could get their Camaros in either Sport Coupe, Berlinetta, or Z28 form. *General Motors 2012*

Below: The Iron Duke four-cylinder engine made its final appearance in 1986, powering over half of the 192,219 Camaros sold that year. *General Motors 2012*

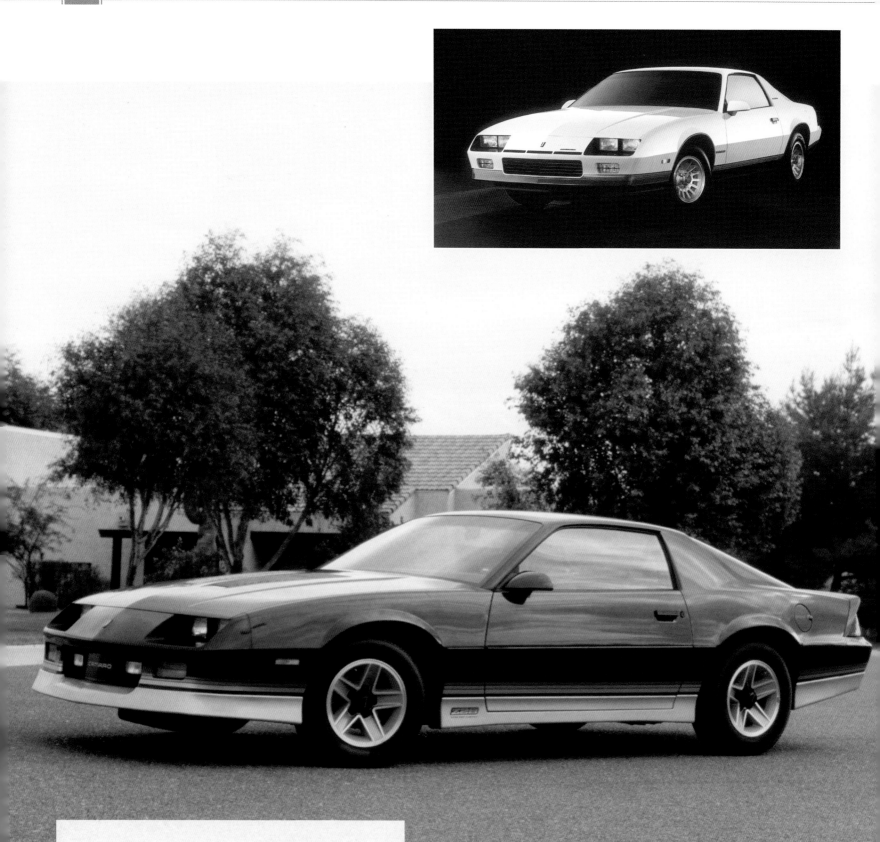

The V-8–powered Z28 proved almost as popular as the four-cylinder base coupe, with 88,132 units sold in 1986. *General Motors 2012*

Inset: Chevrolet sold 99,608 four-cylinder Sport Coupes in 1986. *General Motors 2012*

Above: Chevrolet celebrated the selection of the use of the Camaro in the International Race of Champions (IROC) by offering an IROC-Z appearance package for 1986. In future years, the IROC-Z would become its own model. *General Motors 2012*

1986

MODEL AVAILABILITY	two-door coupe
WHEELBASE	101.1 inches
LENGTH	187.8 inches
WIDTH	72.0 inches
HEIGHT	49.8 inches
WEIGHT	2,912 lbs
PRICE	$9,035
TRACK	60.7/61.5 inches (front/rear)
WHEELS	14 x 7 inches
TIRES	P205/70R14
CONSTRUCTION	unit body
SUSPENSION	modified MacPherson strut, coil springs front/torque arm, struts, coil springs rear
STEERING	recirculating ball
BRAKES	10.5-inch disc front/9.5 x 2.0-inch drum rear
ENGINE	88-horsepower, 2.5-liter I-4; 135-horsepower, 2.8-liter V-6; 155-, 165-, 190-horsepower, 5.0-liter V-8
BORE AND STROKE	4.00 x 3.00 inches (2.5-liter), 3.50 x 2.99 inches (2.8-liter), 3.74 x 3.48 inches (5.0-liter)
COMPRESSION	9.0:1 (2.5-liter) 8.9:1 (2.8-liter), 8.6:1 (155-, 165-horsepower 5.0-liter), 9.5:1 (190-horsepower 5.0-liter)
FUEL DELIVERY	TBI (2.5-liter), MFI (2.8-liter), single four-barrel (155-, 165-, 190-horsepower 5.9-liter), TPI (190-horsepower 5.0-liter)
TRANSMISSION	five-speed manual, four-speed automatic
AXLE RATIO	ranging from 2.73:1 to 3.73:1
PRODUCTION	77,478 V-6, 114,741 V-8

By 1986, the 305-cubic-inch V-8 in the Z28 generated 190 horsepower. *General Motors 2012*

IROC-Z buyers received a detuned version of the Corvette's 350-cubic-inch V-8 that generated 220 horsepower. *General Motors 2012*

In 1987, a Camaro buyer could choose between a base Sport Coupe, Z28, or IROC-Z. *General Motors 2012*

1987

When the 1987 Camaros were revealed, fans of the car found that what Chevrolet took away, Chevrolet could give. A replacement for the Berlinetta appeared: the LT. Not as luxury themed as the Berlinetta, the LT was an option for the Sports Coupe. Some of the features of the LT were body-colored sport mirrors, full wheel covers, a softer ride, a fancy luggage compartment, and auxiliary lighting.

Though power had dropped in 1986, it came back with a vengeance for 1987 as the Corvette's L89 350-cubic-inch V-8 was tapped for optional duty in the Z28 Camaro. Chevy engineers replaced the aluminum heads of the L89 with cast-iron units, and the 'Vette's tubular headers were replaced with cast-iron exhaust manifolds. This was necessary to fit the 5.7-liter V-8 in the Camaro's engine compartment. As far

as power was concerned, the engine, RPO B2L in the Z28, delivered power in droves. Rated at 225 horsepower and costing $1,045, it developed too much torque for the manual transmission in the Camaro. The result was a mandatory option: the automatic transmission. Unfortunately, air conditioning was not available with the 350 engine, as there wasn't enough room in the nose for a big enough radiator to handle all the cooling duties. In spite of these limitations, 12,105 B2Ls were installed IROC-Zs.

The standard Z28 engine continued to be the 5.0-liter V-8, RPO LG4, rated at 170 horsepower. Next up the ladder was RPO L89 which, like the LG4, came with 9.3:1 compression, but the L89, when attached to the automatic transmission, put out 190 ponies. When the same engine was bolted to the five-speed manual, the powerplant delivered 215 horses.

Left: By 1987, the Iron Duke four-cylinder had been mercifully put out of its misery, and the base Camaro Sport Coupe came equipped with 173-cubic-inch V-6. *General Motors 2012*

The top engine option for the 1987 Z28 was a 215-horsepower, 305-cubic-inch V-8. *General Motors 2012*

1987

MODEL AVAILABILITY	two-door coupe, convertible
WHEELBASE	101.1 inches
LENGTH	187.8 inches
WIDTH	72.0 inches
HEIGHT	49.8 inches
WEIGHT	3,062 lbs
PRICE	$10,409
TRACK	60.7/61.5 inches (front/rear)
WHEELS	14 x 6 inches
TIRES	P205/70-14
CONSTRUCTION	unit body
SUSPENSION	modified MacPherson strut, coil springs front/torque arm, struts, coil springs rear
STEERING	recirculating ball
BRAKES	10.5-inch disc front/9.5 x 2.0-inch drum rear
ENGINE	135-horsepower, 2.8-liter V-6; 165-, 170-, 190-, 215-horsepower, 5.0-liter V-8; 225-horsepower, 5.7-liter V-8
BORE AND STROKE	3.50 x 2.99 inches (2.8-liter), 3.74 x 3.48 inches (5.0-liter), 4.00 x 3.48 inches (5.7-liter)
COMPRESSION	8.9:1 (2.8-liter), 8.6:1 (165-, 170-horsepower 5.0-liter), 9.3:1 (190-, 215-horsepower 5.0-liter), 9.0:1 (5.7-liter)
FUEL DELIVERY	MFI (2.8-liter), single four-barrel (165-, 170-horsepower 5.0-liter), TPI (190-, 215-horsepower 5.0-liter, 5.7-liter)
TRANSMISSION	five-speed manual, four-speed automatic
AXLE RATIO	Ranging from 2.72:1 to 3.73:1
PRODUCTION	60,439 V-6, 77,321 V-8

C-Z stuffed Goodyear Eagle P245/50VR rubber inside the wheelwells. In its first year, the Z28 IROC-Z option was popular, with 18,418 units sold. In an effort to "educate" the public, the government mandated that the speedometer top out at 85 miles per hour.

More big news hit the road in 1987: the return of the Camaro convertible, midway through the model year and, in fact, the Camaros were converted to ragtops. Chevrolet contracted American Specialty Company (ASC) to take RPO CC1 T-top-equipped vehicles (T-top cars already had extra bracing installed) and create a full convertible. These were the first convertible Camaros since 1969, and though they commanded a heavy premium ($4,400) they proved very popular. Chevrolet delivered 263 Sport Coupe convertibles, and 744 Z28 and IROC-Z ragtops. Because of the power from the B2L 350 engine, that powerplant was not available in a convertible. Even with the reinforcement from the factory, the high torque threatened to pretzel-ize the structure. But Chevrolet knew that the typical ragtop Camaro buyer wasn't as interested in outright performance as the fresh air experience. Seeing and being seen were, and still are, the primary reasons to embrace an open interior.

In an experiment, Chevrolet built a number of RS Camaros in 1987. Insurance rates on Camaros, especially Z28s, had skyrocketed in certain markets, especially California. The thinking was to equip a 173-cubic-inch V-6 Camaro with some borrowed visual pizzazz from the Z28, including ground effects and spoilers, wheels and tires, and interior, all covered in a monochrome paint scheme, very under the radar. Unfortunately, the car ended up costing about $13,000, less than a grand under a Z28 coupe. And the buyer essentially got a Z28 with a V-6 engine. I'm not saying that it was a poseur-mobile. . . .

In an attempt to reduce confusion when ordering a vehicle, Chevrolet instituted a system of option packages, consisting of a group of normally ordered options bundled together in a "package." In theory, it was a good idea, but invariably, some options were missing that buyers were craving. Gone were the days when a customer would sit with a salesperson and spec out a car to fit their desires exactly. Now it was a matter of take it or leave it. It was a mess.

Total Camaro production dropped to 137,760 for model year 1987 due to another national recession in the wake of the October 19, 1987, stock market crash. Chevrolet shut down the original Camaro plant in Norwood, Ohio, transferring all production to the Van Nuys, California, facility. It had been a big year for the Camaro, yet more changes were coming down the road for 1988.

1988

As the 1988 model year vehicles rolled into showrooms, a couple of models were conspicuous by their absence: the LT and the Z28. Chevrolet was now starting the development process for the next-generation Camaro, and the last thing it wanted to do was devote needed resources to the existing generation. The IROC-Z had proven to be a serious hit, so to simplify matters the Z28 was cut, making the IROC-Z the only performance name in the Camaro camp.

The entire menu for the Camaro consisted of two models of the Sports Coupe: a closed car and the convertible, and the IROC-Z, also available in coupe and ragtop form. Engine choices were a little confusing, but it went like this: the base engine on the Sports Coupe was the 173-cubic-inch, 135-horsepower V-6; and the 5.0-liter, 170-horse V-8 was an option in the coupe. For buyers choosing a Sport Coupe Convertible, the 5.0-liter V-8 was a no-cost option. The IROC-Z Coupe and Convertible came standard with the same 5.0-liter, 170-horsepower V-8; except when installed in the IROC-Z, the carburetor was replaced by throttle body fuel injection mounted on the same cast-iron intake manifold.

The only optional engine for the IROC-Z was RPO B2L, the 350-cubic-inch V-8, now rated at 230 horsepower. Except buyers couldn't get it in the ragtop—it was a coupe engine only, with a mandatory automatic transmission. Got all that? Factor all that, along with a long laundry list of confusing "Option Packages," and mix in a struggling economy, and it's a wonder Chevrolet sold 96,275 Camaros in model year 1988. What is surprising is that 3,761 IROC-Z convertibles (base price $18,015) were sold. That was serious money.

Above: In 1988, the Z28 once again disappeared from the lineup; the IROC-Z supplanted it completely. *General Motors 2012*

Left: The convertible option proved popular among Camaro buyers. *General Motors 2012*

Below: By 1988, the science of aerodynamics had progressed dramatically, and the Camaro's spoilers served as more than mere ornamentation. *General Motors 2012*

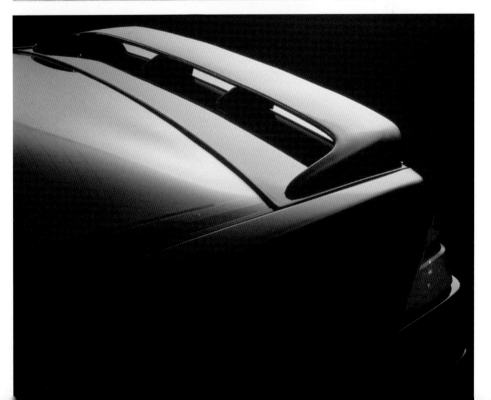

Above: While the IROC-Z got all the press, the Sport Coupe still generated the lion's share of sales volume. *General Motors 2012*

Right: The angular cockpit designs en vogue in the 1980s haven't aged well, but at the time, they were all the rage. *General Motors 2012*

1988

MODEL AVAILABILITY	two-door coupe, convertible
WHEELBASE	101 inches
LENGTH	192.0 inches
WIDTH	72.8 inches
HEIGHT	50.3 inches
WEIGHT	3,054 lbs
PRICE	$10,995
TRACK	60.7/61.6 inches (front/rear)
WHEELS	15 x 7 inches
TIRES	P215/65R15
CONSTRUCTION	unit body
SUSPENSION	modified MacPherson strut, coil springs front/torque arm, struts, coil springs rear
STEERING	recirculating ball
BRAKES	10.5-inch disc front/9.5 x 2.0-inch drum rear
ENGINE	160-horsepower, 2.8-liter V-6; 170-, 195-, 220-horsepower, 5.0-liter V-8; 230-horsepower, 5.7-liter V-8
BORE AND STROKE	3.50 x 2.99 inches (2.8-liter), 3.74 x 3.48 inches (5.0-liter), 4.00 x 3.48 inches (5.7-liter)
COMPRESSION	8.9:1 (2.8-liter), 9.3:1 (5.0-, 5.7-liter)
FUEL DELIVERY	MFI (2.8-liter), TBI (5.0-, 5.7-liter)
TRANSMISSION	five-speed manual, four-speed automatic
AXLE RATIO	Ranging from 2.73:1 to 3.45:1
PRODUCTION	42,820 V-6, 53,455 V-8

Above: For the 1988 model year, the Z28 IROC changed induction systems; gone was the venerable carburetor, replaced by throttle-body fuel injection. There were a pair of 5.0-liter V-8s on the roster, but the more powerful one could only be had with an automatic transmission.

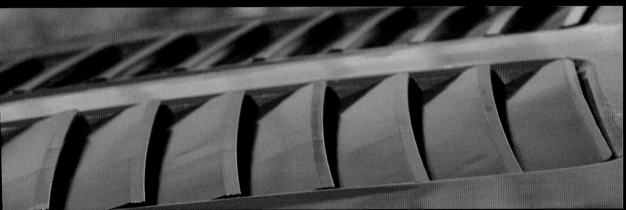

Left: The hood louvers on the 1988 Z28 IROC were nonfunctional. Stylists used them to differentiate the Z28 from the "regular" Camaro in lieu of paint stripes. The vents tended to collect water and debris, and cleaning them was a pain in the butt.

Right: Lower body cladding gave the appearance of a race car's ground effect package. In truth, on a street car, the aerodynamic improvement from such bolt-ons was questionable, but they did make a fine place to mount identification badging.

Above: One of many Z28 IROC badges found inside and out of the 1988 model. This particular one is located at the far right side of the dashboard, constantly reminding the front passenger just what they were in. Leave it to Chevrolet to address this chronic problem.

The lower air dam and rocker panel extensions were aerodynamic aids that were more effective at the race speed that actual IROC cars carried. But there was no denying that the added cladding created a more balanced visual package when mounted on a 1989 IROC-Z.

Above: The lines of the third-generation Camaro lent themselves to a convertible treatment. The convertible "stack," or folded top, was engineered to be completely hidden when down, and that feature helped give the Camaro a sleek profile.

1989

For 1989, the deck was shuffled again. Gone was the Sport Coupe, replaced by the RS. The popularity of the RS package spurred Chevrolet to use it as the entry-level model. It was equipped with IROC-Z ground effect pieces and handsome aluminum wheels made to resemble the rolling stock from the performance model, yet it packed, in base form, the 173-cubic-inch V-6 from years prior, complete with 135 horsepower. The only optional engine in the RS was the 5.0-liter V-8, rated at 170 horsepower in this application. This same powerplant was standard in the RS convertible, IROC-Z coupe and IROC-Z ragtop. Optional engines in the IROC-Z consisted of the LB9 version of the 5.0-liter V-8. Its output varied, depending on transmission choice and exhaust system configuration. With an automatic transmission, the engine put out 195 horses. Install a five-speed manual transmission and single exhaust, and you had

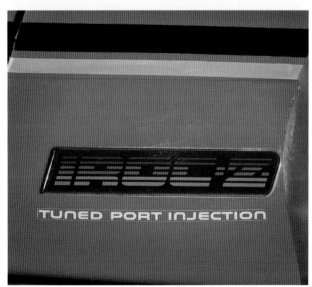

Above: Tuned port injection made its debut on Chevrolet vehicles in the Corvette, and it didn't take long for the induction system to make its way onto the Camaro. This 1989 IROC-Z enjoys all-weather tractability, superior fuel mileage, and reduced emissions from its carbureted predecessors.

220 ponies turning the rear tires. Slip a dual exhaust beneath the car, in conjunction with the manual gearbox, and output rose to 230. For even more power, the IROC-Z coupe only could be outfitted with the L98, 350-cubic-inch V-8. When a single exhaust system was used, the big engine cranked out 230-horsepower. When fitted with the dual exhaust, it was rated at 240. But the L98 was only available with an automatic transmission bolted to the back of the block.

Chevrolet, long involved in racing, both directly and indirectly, was approached to provide race cars for the SCCA's Showroom Stock racing series. Actually, the series started in Canada as the Canadian Players Challenge and eventually included SCCA and IMSA (International Motor Sports Association) events in America. Many saw this as an opportunity for the Camaro to excel. But when initial tests

were run, the racetrack, as it often does, revealed weaknesses in the vehicle that would rarely if ever crop up in street driving.

The biggest hurdle involved the brakes. The stock IROC-Z binders were woefully inadequate, but by clever parts picking in the GM bins, the brake system was made race ready. But then another problem raised its head; during hard deceleration, fuel starvation would occur. A small 0.5-gallon reservoir was built into the fuel system to prevent it. And so it went, as the test drivers would uncover a problem, and the engineers worked to resolve it.

By the time they were finished, a true race-ready Camaro was rolling off the assembly line in Van Nuys, California. But this was not a package that was common knowledge to the public. In fact, most dealers had no idea that the 1LE package even existed. And if buyers didn't know how to order one,

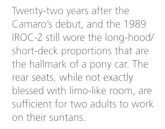

Twenty-two years after the Camaro's debut, and the 1989 IROC-Z still wore the long-hood/short-deck proportions that are the hallmark of a pony car. The rear seats, while not exactly blessed with limo-like room, are sufficient for two adults to work on their suntans.

in the proper sequence, they would come up empty-handed. Here's how it worked:

First, you had to order an IROC-Z coupe with the RPO G92 Performance Axle, in conjunction with an order to *delete* C60 air conditioning. As the vast number of buyers sprang for A/C, they were automatically out of 1LE consideration. Engine choices were easy: either the 5.7-liter V-8 with the automatic transmission, or the 5.0-liter mill attached to the five-speed manual. By ordering the G92 option, the car was equipped with four-wheel disc brakes, dual catalytic converter exhausts, an engine oil cooler, P245/50AR16 Goodyear Eagle tires, a 5,500-rpm tachometer, and a 145-miles per hour speedometer. By not ordering air

Above: Handsome five-spoke aluminum wheels graced every 1989 Camaro IROC-Z that rolled off the line. With four-wheel disc brakes, a sport suspension, and serious rolling stock, this iteration of the Chevrolet F-body didn't embarrass itself when the pedal hit the metal. A 230 horsepower, 350-cubic-inch V-8 was available on IROC-Z coupes.

Above: Chevrolet built 3,940 IROC-Z Camaros in 1989, with a list price of $18,945. This was the first year for the "passkey" system, which used a resistor pellet in the key in an effort to reduce vehicle theft. A computer in the car read the coded resistor, and if it matched, the computer would allow the car to start.

1989

MODEL AVAILABILITY	two-door coupe, convertible
WHEELBASE	101.1 inches
LENGTH	192 inches
WIDTH	72.8 inches
HEIGHT	50.3 inches
WEIGHT	3,082 lbs
PRICE	$11,495
TRACK	60.7/61.6 inches (front/rear)
WHEELS	15 x 7 inches
TIRES	P215/65R15
CONSTRUCTION	unit body
SUSPENSION	modified MacPherson strut, coil springs front/torque arm, struts, coil springs rear
STEERING	recirculating ball
BRAKES	10.5-inch disc front/9.5 x 2.0-inch drum rear
ENGINE	135-horsepower, 2.8-liter V-6; 170-, 195-, 220-, 230-horsepower, 5.0-liter V-8; 230-, 240-horsepower, 5.7-liter V-8
BORE AND STROKE	3.50 x 2.99 inches (2.8-liter), 3.74 x 3.48 inches (5.0-liter), 4.00 x 3.48 inches (5.7-liter)
COMPRESSION	8.9:1 (2.8-liter), 9.3:1 (5.0-, 5.7-liter)
FUEL DELIVERY	MFI (2.8-liter), TPI (5.0-, 5.7-liter)
TRANSMISSION	five-speed manual, four-speed automatic
AXLE RATIO	Ranging from 2.73:1 to 3.45:1
PRODUCTION	42,729 V-6, 68,010 V-8

conditioning, the 1LE package was triggered, and this added heavier-duty front disc brakes and calipers, special front and rear shock absorbers, an aluminum driveshaft, fuel tank baffle, and special Durometer jounce bumpers. The result of this nearly $700 option was a very serious race car that could actually be driven on the street.

In actuality, the option became available late in the 1988 model year, but only four units were built. But it didn't take long for the racing community to get the word out, and in 1989 Chevrolet built 111 of these rockets. Was the car worth the effort? The Camaro 1LE won every race in the 1989 SCCA Escort Endurance Championship Series, as well as taking the "Car of the Year" award in IMSA's Firestone Firehawk series.

Theft, long a problem with performance automobiles, was an issue that manufacturers had done little to prevent for many years. But starting in 1986 on the Corvette, Chevrolet

By 1990, the RS was no longer an appearance package; it had replaced the Sport Coupe as the base Camaro. *General Motors 2012*

introduced a passive theft-prevention measure, with the fitting of a coded resistor to the ignition key. A sensor in the column assembly would "read" the resistor, and if it matched the data in the computer, the car would be allowed to start. With the introduction of this system, Corvette thefts had plummeted. This technology trickled down to the Camaro in the 1989 model year.

As model year 1989 drew to a close, the Camaro saw an uptick in total sales to 110,739 units. This was an improvement for a car that had seen very little change from the previous year. For a vehicle in a holding pattern until its successor was finished being developed, the Camaro was showing great legs in the marketplace. Yet Chevrolet had a few more tricks up its sleeve for 1990.

1990

Chevrolet had contracted with the International Race of Champions organization to provide vehicles to the race

series through the end of 1989, which meant that Camaro IROC-Zs couldn't be built after midnight, December 31, 1989. This had the potential to leave Chevrolet in an uncomfortable position. Without the IROC-Z in the lineup, what would the Camaro offer as its high-performance vehicle? Many in Chevrolet felt that bringing back the Z28 name mid way through the model year would confuse people.

Jim Perkins, Chevrolet's General Manager, decided to stop production of all 1990 Camaros on December 31, 1989, and release the 1991 Camaro in March 1990. As a result of this decision, production of the 1990 Camaro was abbreviated and the sales numbers reflected this. Total Camaro sales for the 1990 model year were only 34,986 units, with the vast majority being RS coupes, with 28,750 built. The 1LE option was still available, and Chevrolet assembled 62 of the road-legal rockets.

Yet for a model that was inching closer to its generational end, the 1990 Camaro introduced a few surprises. The base

By 1990, the IROC-Z generated a respectable 245 horsepower, thanks to a 350-cubic-inch engine sourced from the Corvette. *General Motors 2012*

By 1990, Camaro sales had plummeted to 34,986 units, roughly 2,000 of which were convertibles. *General Motors 2012*

1990

MODEL AVAILABILITY	two-door coupe, convertible
WHEELBASE	101.1 inches
LENGTH	192 inches
WIDTH	72.8 inches
HEIGHT	50.4 inches
WEIGHT	2,975 lbs
PRICE	$10,995
TRACK	60.7/61.6 inches (front/rear)
WHEELS	15 x 7 inches
TIRES	P215/65R15
CONSTRUCTION	unit body
SUSPENSION	modified MacPherson strut, coil springs front/torque arm, struts, coil springs rear
STEERING	recirculating ball
BRAKES	10.5-inch disc front/9.5 x 2.0-inch drum rear
ENGINE	140-horsepower, 3.1-liter V-6; 170-, 210-, 230-horsepower, 5.0-liter V-8; 245-horsepower, 5.7-liter V-8
BORE AND STROKE	3.50 x 3.31 inches (3.1-liter), 3.74 x 3.48 inches (5.0-liter), 4.00 x 3.48 (5.7-liter)
COMPRESSION	8.5:1 (3.1-liter), 9.3:1 (5.0-, 5.7-liter)
FUEL DELIVERY	MFI (3.1-liter), TPI (5.0-, 5.7-liter)
TRANSMISSION	five-speed manual, four-speed automatic
AXLE RATIO	Ranging from 2.73:1 to 3.45:1
PRODUCTION	12,743 V-6, 22,243 V-8

173-cubic-inch (2.8-liter) V-6, standard on the RS coupe, was enlarged to 191-cubes (3.1-liter). With the increased displacement, power rose as well, from 135-horsepower to 140. At the opposite end of the engine menu, the 350-cubic-inch (5.7-liter) RPO L89 V-8 received lighter pistons, which bumped output up from 240-horsepower to 245.

The federal government had decreed that starting September 1, 1989, all cars built after that date must have a passive restraint mechanism to improve driver safety. Chevrolet addressed the law by fitting the Camaro with a supplemental inflatable restraint (SIR), also known as an air bag. This feature was installed in the steering wheel hub.

1991

With the shortened model year, all eyes were on the 1991 model year Camaros as the Z28 made its return to the lineup, replacing the IROC-Z. Changes to the exterior were especially pronounced on the Z28, with its restyled lower

Right: For 1991, the RS was once again the base model Camaro. *General Motors 2012*

Below: When the International Race of Champions quit using Camaros, Chevrolet lost the right to use the IROC name, leading to a second rebirth of the Z28. *General Motors 2012*

The Z28 retained the 245-horsepower L98 engine used in the last IROC-Z, making the 1991 Z28 the fastest Camaro in 20 years. *General Motors 2012*

body trim pieces and the very tall rear spoiler. For the first time, the RS was available with 16-inch aluminum wheels. In the beginning of the model year, the RS coupe, as in years before, used the V-6 as the standard engine, while the RS convertible employed the 5.0-liter V-8 as the base engine. Partway through the model year, the RS ragtop's base engine became the same 3.1-liter V-6 found in the RS coupe. The standard engine in both the Z28 coupe and convertible was the 5.0-liter V-8, rated at 205 horsepower.

Rear axle ratios for use with that engine stressed fuel economy rather than neck-snapping acceleration: 3.08:1 with the five-speed manual transmission, 2.73:1 with the automatic transmission. Still on tap for speed demons was the 1LE option package, which 478 buyers signed up for. Air

conditioning, or rather the lack of it, was still a prerequisite for the track-ready setup. As in prior years, the fog lamps were deleted in the name of weight savings, as well as improving airflow to the radiator.

Another entry to the Camaro performance realm hit the street for 1991: RPO B4C. This option was not advertised to the public, as it was aimed at law enforcement agencies. It was an interesting mix of components not normally assembled in a Camaro. The RS coupe was stuffed with either the 5.0- or 5.7-liter TPI Z28 engines, complemented by the Z28's suspension. Rolling stock consisted of P245/50-ZR16 tires mounted on 16-inch wheels, while an engine oil cooler helped to keep the powerplant cool during the long periods waiting for a perp to flash by. Rear disc brakes were installed, as

Camaro sales skyrocketed in 1991, and the RS coupe led the charge.
General Motors 2012

well as a limited slip differential. Partway through the model year, the very heavy-duty front brakes from the 1LE package became an option on the police interceptor. Unlike the regular 1LE package available to civilians, air conditioning was fitted to the B4C vehicles. By the time the model year came to a close, 592 of the very capable cruisers had been built. It was not an inexpensive option: $3,135 for 5.0-liter-equipped cars, $3,950 with the 5.7-liter engine beneath the long hood. All the better for collaring social deviants.

As the extended model year drew to a close, 1991 proved to be a very good year in CamaroLand, as total sales added up to 100,838. When you consider that the public was aware that a new Camaro was just over the horizon, it was clear that customers were still in love with the Chevy F-body. While it was interior-space challenged, and not exactly inexpensive, it was one of the finest-handling vehicles available in America, and it delivered more bang for the buck than anything else. It was what is called Good Value.

1991

MODEL AVAILABILITY	two-door coupe, convertible
WHEELBASE	101.1 inches
LENGTH	192 inches
WIDTH	72.4 inches
HEIGHT	50.4 inches
WEIGHT	3,103 lbs
PRICE	$12,180
TRACK	60.7/61.6 inches (front/rear)
WHEELS	15 x 7 inches
TIRES	P215/65R15
CONSTRUCTION	unit body
SUSPENSION	modified MacPherson strut, coil springs front/torque arm, struts, coil springs rear
STEERING	recirculating ball
BRAKES	10.5-inch disc front/9.5 x 2.0-inch drum rear
ENGINE	140-horsepower, 3.1-liter V-6; 170-, 205-, 230-horsepower, 5.0-liter V-8; 245-horsepower, 5.7-liter V-8
BORE AND STROKE	3.50 x 3.31 inches (3.1-liter), 3.74 x 3.48 inches (5.0-liter), 4.00 x 3.48 inches (5.7-inches)
COMPRESSION	8.5:1 (3.1-liter), 9.3:1 (5.0-, 5.7-liter)
FUEL DELIVERY	MFI (3.1-liter), TPI (5.0-, 5.7-liter)
TRANSMISSION	five-speed manual, four-speed automatic
AXLE RATIO	Ranging from 2.73:1 to 3.42:1
PRODUCTION	31,722 V-6, 69,116 V-8

1992

The last year of third-generation production was also the 25th anniversary of the Camaro; and Chevrolet, never one to let a milestone pass without a profit, unveiled an exterior appearance option, RPO Z03, the Heritage Package. For $175, buyers went home with a Camaro festooned with hood and deck stripes, black headlight recesses, a body-color grille, and a unique deck lid badge. For Z28 customers wanting the Heritage Package, their cars had to be one of four colors: black, Arctic White, Purple Haze, or Bright Red. If you sprang for an RS instead, you could acquire the Heritage Package in any of the above colors, as well as Polo Green II Metallic.

The rest of the news for the 1992 Camaro was essentially to make it to the end of the model year with decent sales results. Chevy was pleased when the year's tally came in at 70,007 units, including 705 1LEs, and 589 B4C police packages.

But the end of third-generation production also marked the end of Camaro production at the Van Nuys, California, plant. Starting with the 1993 Camaro, the F-body would be built north of the border, Ste. Therese, Quebec, Canada, to be exact. The third-generation Camaro had been in production 11 years; that is an eternity in the life cycle of an automobile. Through constant improvement and attention to the needs and wants of the customer, the Camaro moved into its next iteration stronger than ever, with an enthusiast base hungry for the next version. Chevrolet didn't disappoint them.

Above: While Chevrolet offered convertible versions of the Camaro, it retained the optional T-tops. *General Motors 2012*

Left: The third-generation Camaros matured into serious performance cars over their eleven-model-year production run. *General Motors 2012*

Left: Purists might howl about Z28 stripes on base RS models, but it's hard to argue with the looks of the resulting cars. *General Motors 2012*

Opposite top: Since the third-generation Camaro wasn't originally designed to be a convertible, chassis rigidity suffered when the top was removed. *General Motors 2012*

Opposite bottom: The last third-generation Z28s generated 245 horsepower; that figure would take a quantum leap forward with the introduction of the fourth-generation Camaros the following year. *General Motors 2012*

Below: Total convertible production equaled roughly 10 percent of total Camaro output in 1992. *General Motors 2012*

Above: The world's fastest Camaro started life as a 1982; it has a 1985 nose, as that nose is more aerodynamic. A full interior is still in the car, and it currently holds 14 world records at the Bonneville Salt Flats, with a top speed of 251 miles per hour using a 448-cubic-inch normally aspirated V-8.

Opposite: The engine of the 757 Camaro uses headers that dump exhaust ahead of the front wheels. The fuel injectors were made by Tim Engler and are similar to Hilborn units. The stacks are big enough to drop a hardball in. An airbox sits atop the engine when it's running the salt, using the high-pressure zone that builds up at the base of the windshield at speed.

1992

MODEL AVAILABILITY	two-door coupe, convertible
WHEELBASE	101.1 inches
LENGTH	192 inches
WIDTH	72.4 inches
HEIGHT	50.4 inches
WEIGHT	3,103 lbs
PRICE	$12,075
TRACK	60.7/61.6 inches (front/rear)
WHEELS	15 x 7 inches
TIRES	P215/65R15
CONSTRUCTION	unit body
SUSPENSION	modified MacPherson strut, coil springs front/torque arm, struts, coil springs rear
STEERING	recirculating ball
BRAKES	10.5-inch disc front/9.5 x 2.0-inch drum rear
ENGINE	140-horsepower, 3.1-liter V-6; 170-, 205-, 230-horsepower, 5.0-liter V-8; 245-horsepower, 5.7-liter V-8
BORE AND STROKE	3.50 x 3.31 inches (3.1-liter), 3.74 x 3.48 inches (5.0-liter), 4.00 x 3.48 inches (5.7-liter)
COMPRESSION	8.5:1 (3.1-liter), 9.3:1 (5.0-, 5.7-liter)
FUEL DELIVERY	MFI (3.1-liter), TPI (5.0-, 5.7-liter)
TRANSMISSION	five-speed manual, four-speed automatic
AXLE RATIO	Ranging from 2.77:1 to 3.73:1
PRODUCTION	23,825 V-6, 46,182 V-8

INDIANAPOLIS
500
™
THE SEVENTY SEVENTH • MAY 30, 1993

Z28

GENERATION FOUR

1993-2002 Real Performance, Real Style

As the Camaro entered its 26th year of production, few realized how close the Camaro came to becoming a front-drive crap box. In a nod to history, Ford Motor Company was influential for the Camaro continuing as a front-engine, rear-drive muscle car. General Motors was developing a front-drive Camaro under the program "GM80," and with the debut of the Saturn line, gutless, plastic-bodied, front-drive small cars were all the rage. The imports were unloading millions of soulless transportation devices into America, and bean-counters at GM, who wouldn't know a performance car if it ran over them, were enamored with the thought of huge profits on a common front-drive platform.

Ford Motor Company went down that road, inflicting the Probe upon the public. When that vehicle was first announced, it was framed as the next-generation Mustang. The Mustang faithful raised such a huge outcry, Ford backed down and left the Mustang as a front-engine, rear-drive sporty car. In the meantime, the Probe generated poor sales numbers. Chevrolet watched

Subtle it ain't, but the door graphics on the 1993 Camaro Z28 Indy 500 replica let everyone know that the forth generation had made another trip to Indiana in May. The Camaro has been a popular pace car at the Indianapolis 500 over the years, as well as at other venues, such as NASCAR events.

197

Debuting at the 1989 Los Angeles International Auto Show, the California Camaro was designed at General Motors' Advanced Concepts Center in Los Angeles. It showcased a number of styling cues that would surface on the fourth-generation Camaro.

Right: The California Camaro concept vehicle had an aggressively styled nose, too aggressive for production. But like most concept vehicles, it went over the line enough for the production version to pull back from the concept's excesses to reach reality.

Left: The scissor-hinge doors opened to reveal a futuristic interior, with a swiveling driver's seat that was different than the passenger's and a movable steering wheel and pedals. Designer Chuck Jordan led a team of 50 stylists to create a vehicle that stopped people in their tracks. They succeeded.

Below: Unlike so many concept vehicles, the California Camaro was built out of metal, not fiberglass. The vehicle was constructed in just six months, far faster than most concept vehicles' gestation time. The vehicle was only 186.4 inches long, compared with production cars at 192 inches.

Above: The elongated nose with the long overhang defined the look of the fourth-generation Camaro. *General Motors 2012*

Below: Early fourth-generation drawings hint at the look the Camaro would adopt toward the end of its original production run. *General Motors 2012*

Chevrolet was quick to assure everyone that a ragtop would be coming; but for 1993, a closed car was the only choice. Design work on the convertible was extensive, and it just hadn't been finished in time to make the 1993 product launch.

But what a choice! The new body was the picture of slippery, with a radically sloped windshield, raked back 68 degrees. The base of the windshield extended so far forward, part of the engine was under the leading edge of the glass. The back window was nearly horizontal as well, and at the rear, a graceful spoiler wrapped around the end of the car. One of the plans for the axed GM80 program was the use of plastic body panels. That feature did move forward into production, to the point that the only steel body panels on the new Camaro were the hood and rear quarter panels. Sheet molded compound (SMC) was a material made up of chopped fiberglass and polyester resin. It had been developed by Fisher Body Division of General Motors, and it was used to make the doors, hatch, roof, and spoiler assembly. The front fenders and fascia were made using the reaction injection molded (RIM) method, while the rear fascia was constructed of reinforced polyurethane. Rust wouldn't be a problem.

Beneath the swoopy bodywork was a platform that was essentially carried over from 1992, but tuning of the structure resulted in a 23 percent stiffer car. The wheelbase was

this drama unfold, and when it came time to finalize the drivetrain configuration of the next-generation Camaro, they came down firmly in the front-engine, rear-drive camp. If there was one thing Chevrolet knew, it was the desires of Camaro buyers.

So when the newest version of the Camaro was unveiled, there was a bit of a surprise: there was no convertible. Only two models were available: the base Coupe and the Z28.

virtually unchanged at 101.1 inches, an increase from 1992 of 0.1 inch. Overall, the car had grown: 0.5 inches longer, 0.9 inches taller, and 1.7 inches wider. Weight was up too, as the base 1993 Camaro tipped the scales at 3,241 pounds, up from the 1992 RS's 3,060 pounds. Yet the suspension, especially at the front, was an improvement.

The strut setup used in the third generation was ditched, replaced by a short/long (control) arm (SLA) that worked wonderfully. Rack-and-pinion steering was finally used, giving the Camaro a much crisper steering feel. The rear suspension was, for the most part, carried over from the prior year. Consisting of a live rear axle, it utilized a multilink

arrangement, using a pair of trailing lower control arms, a tie rod, tie rod brace, and a torque arm. Both front and rear suspension used high-gas-pressure de Carbon shock absorbers. The base Camaro was fitted with disc brakes in the front, drum brakes at the stern. Anti-lock brakes (ABS) were standard on all Camaros. The Z28 employed four-wheel disc brakes, with 10.9-inch rotors in the front and 11.4-inch rotors at the rear.

Powerplants in the new Camaro were pretty straight-forward. The base coupe was equipped with the RPO L32 3.4-liter V-6. This engine was advertised with 160 horsepower and 200 lb-ft of torque. Induction duties were handled by

It's hard to believe, but a 3.1-liter all-aluminum V-6 is beneath the steeply sloped hood. Though the 1989 California Camaro concept vehicle looks like it would be a mid-engine design, it was a front-engine, rear-drive automobile. The exterior mirrors would be used on the fourth-generation Camaro virtually unchanged.

Above: The LT1 engine was originally installed in the Corvette, and when the fourth-generation Camaro Z28 hit the street, the able powerplant was slipped under the sleek hood of the F-body. It boasted a 10.3:1 compression ratio, requiring premium unleaded fuel. The gas tank was smallish, at 15.5 gallons capacity, but you probably needed to stretch your legs anyway.

Right: Vivid outside, vivid inside. The 1993 Camaro Z28 Indy pace car carried the two-tone, swirling bands of color motif into the interior. The winner of the Indianapolis 500 in 1993 was Emerson Fittipaldi, who took one of the two actual pace cars home as a prize.

sequential port fuel injection (SFI), which was similar to the system developed for the LT1 V-8 engine. This efficient design delivered a healthy dose of performance with commendable fuel mileage: 19 mpg in the city, 28 mpg on the highway. Those figures were good with either the five-speed manual transmission that was standard in the coupe or the optional 4L60 four-speed automatic transmission.

The Z28 for 1993 enjoyed a step up the horsepower ladder in the form of the Corvette's LT1 V-8. While this engine displaced 350 cubic inches like the year before, it was significantly different, enough to earn it the name Gen II. Major changes had been lavished on the "new" engine, and they included a revision of the induction system. The tubes, induction plenum, and intake manifold that had been atop the engine before were replaced by a one-piece cast-aluminum intake manifold. This component not only improved flow into the heads, it lowered the height of the engine 3.5 inches,

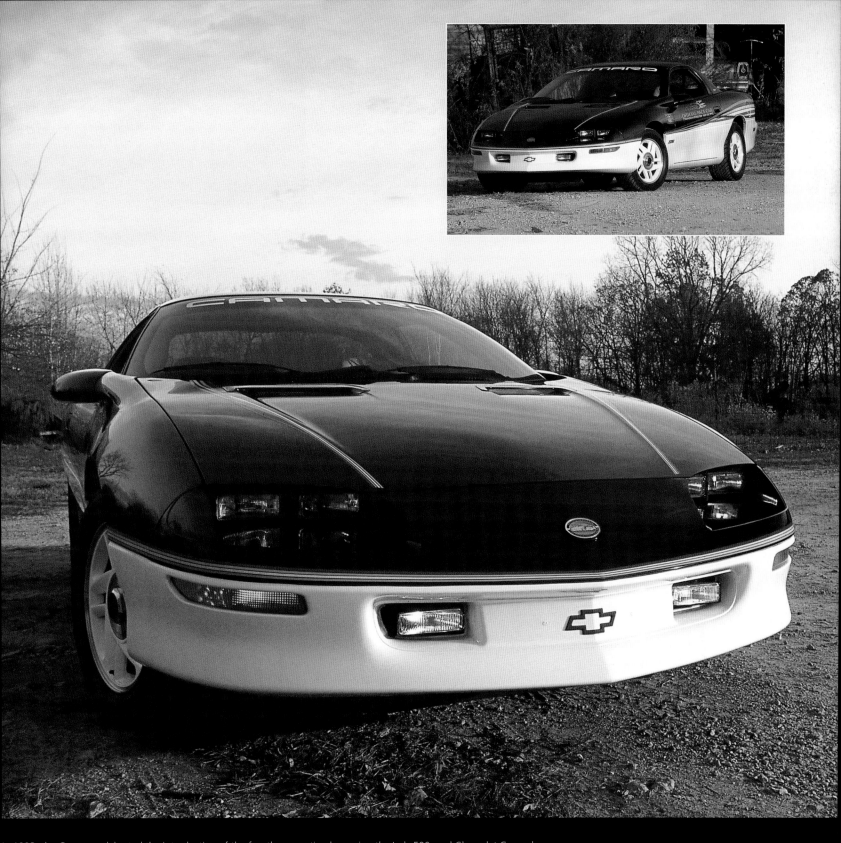

In 1993, the Camaro celebrated the introduction of the fourth generation by pacing the Indy 500, and Chevrolet General Manager Jim Perkins was behind the wheel of the actual pace car on race day. Very few modifications from stock were needed to keep the Z28 in front of the pack. The car tipped the scales at 3,373 pounds, and the LT1 5.7-liter engine would get the slippery F-body to 60 miles per hour in just 6.3 seconds. If a drag strip were in front of the car, the finish line would pass into the vehicle's wake in 14.8 seconds at 95.5 miles per hour, fast enough to get to the race on time.

1993

MODEL AVAILABILITY	two-door coupe
WHEELBASE	101.1 inches
LENGTH	193.2 inches
WIDTH	74.1 inches
HEIGHT	51.3 inches
WEIGHT	3,330 lbs
PRICE	$13,399
TRACK	60.7/60.6 inches (front/rear)
WHEELS	16 x 8 inches
TIRES	P235/55R16
CONSTRUCTION	unit body
SUSPENSION	upper & lower control arms, coil springs front/live axle, lower trailing arms, coil springs rear
STEERING	rack and pinion
BRAKES	10.9-inch disc front/9.5 x 2.0-inch drum rear
ENGINE	160-horsepower, 3.4-liter V-6; 275-horsepower, 5.7-liter V-8
BORE AND STROKE	3.62 x 3.31 inches (3.4-liter), 4.00 x 3.48 inches (5.7-liter)
COMPRESSION	9.0:1 (3.4-liter), 10.5:1 (5.7-liter)
FUEL DELIVERY	SFI (3.4-liter), MFI (5.7-liter)
TRANSMISSION	five- and six-speed manual, four-speed automatic
AXLE RATIO	2.73:1, 3.42:1
PRODUCTION	21,253 Base, 17,850 Z28

Each 1993 Indianapolis 500 pace car replica wore a special badge on the nose commemorating the event. Dimples in the bumper cover were for the mounting of front license plates in states that required them. With its smooth nose, the fourth-generation Camaro was strictly a bottom-breather.

a huge amount that was used to advantage by the stylists. Working with the new intake were AC Rochester Multec fuel injectors, fed by a fuel rail flanking the intake manifold. The Z28 used a different manual transmission than its six-cylinder counterpart; a Borg-Warner T56 box handled the ratios. This tranny was a tough unit, hardy enough to be installed in the Dodge Viper.

Another innovation for the LT1 engine was its reverse cooling system. In order to keep cylinder head temperatures low, coolant would flow through the heads before being directed into the block. This flow path generated uniform cylinder bore temperatures, which reduced ring bore friction, improving mileage. Firing the air/fuel mixture was a high energy ignition (HEI) system that worked with the vehicle's powertrain control module to send the power to the spark plug at precisely the right moment to maximize output and minimize emissions.

The camshaft and exhaust systems were massaged as well, helping the LT1 deliver a consistent 275 horsepower at 5,000 rpm, and 326 lb-ft of twist at just 2,400 rpm. The rear tires were in jeopardy. The Z28 would cover the quarter-mile in 14.7 seconds, tripping the lights at 97 miles per hour. Its top speed was 150 miles per hour. Serious performance was back in the Camaro camp.

The 1LE package soldiered on, but without an aluminum driveshaft. The suspension bushings were firmed up, heavier-duty de Carbon shocks were installed, and the stabilizer bars were enlarged. Yet again, air conditioning and the 1LE package were not to meet. Nineteen 1LEs were built, the package stickering for $310.

Chevrolet hadn't ignored the interior of the new Camaro. It was as swoopy as the exterior. High-backed bucket seats employed beefy side bolsters to hold the driver and front passenger in place during transitional maneuvers. The dashboard and instrument panel were more curvaceous and organic than the preceding year, and for the first time, the front passenger enjoyed the security of an air bag. The air conditioning system, in an attempt to help the environment, was charged with R134a refrigerant, a Camaro first.

As this was the debut year for the latest incarnation of the Camaro, Chevrolet's 2+2 was tapped to once again pace a little race held each May in Indianapolis, Indiana. This was the 77th running of the Indianapolis 500-Mile Race, and Chevrolet Manager Jim Perkins was behind the wheel of the vivid two-tone Pace Car. Equipped with the standard LT1 5.7-liter V-8, it had no problem staying in front of the snarling race cars. Eventual race winner Emerson Fittipaldi was given one of the two actual Pace Cars as a prize for capturing the checkered flag. Of course, what good is pacing the Indy 500 if you can't sell Pace Car replicas? Chevrolet did exactly that, pricing the option at $995; and when the dust had settled, 633 had been sold to the public

Overall sales in the fourth-generation's freshman year were low, coming in at 39,103. This was due to teething problems at the Camaro plant, a refurbished facility in Ste. Therese, Quebec, Canada. Some people within Chevrolet were worried that Americans might not respond favorably to a Canadian-built American muscle car, but focus groups quickly put that concern to rest. As the fourth-generation Camaro watched the 1990s pass, sales would rise and fall, but the country of the vehicle's assembly would never be an issue.

Right above: The flow of the sheet metal continued to the rear, where a graceful spoiler spanned the huge rear window. Instead of a stylistically clumsy housing for the center-mounted stoplight, the Camaro's designers placed the light within the rear spoiler without marring the overall design. The two-tone paint on the 1993 Indy 500 pace car was particularly striking.

Right: The nose of the 1993 Camaro Z28 Indy 500 pace car is the road-going version of a shark on wheels. Within the long prow of the Camaro is the radiator and 5-miles-per-hour bumper. Side marker lights were smoothly integrated into the curvaceous design. The downside to the sleek, ground-hugging front end was the challenger of parking.

Left: These white-finished wheels were unique to the 1993 Camaro Indianapolis 500 pace car replicas. Equipped with a Corvette-sourced LT1, 5.7-liter V-8, the $18,210 vehicle could spin the wheels fast enough to cover the quarter-mile in 14.7 seconds at 97 miles per hour. The top speed of the slippery car was 150 miles per hour.

The 350-cubic-inch engine in the Z-28 pumped out 275 horsepower, the most of any small-block Camaro since the original LT-1 of 1970. *General Motors 2012*

1994

The Camaro entered the 1994 model year firing on all cylinders, as announcements of coming attractions brought the Camaro faithful to the showroom. The big news for 1994 was the return of a convertible Camaro. Built on the same production line as the coupe in Canada, it was blessed with heavier-gauge steel in parts of the unibody structure, as well as other reinforcements. The power top was fully lined and included a heated glass rear window, a luxury touch. With a

price of $18,475 for the base convertible, and $22,075 for the Z28 ragtop, a bit of luxury was expected. But the 1994 Camaro, especially in Z28 trim, could run with the best.

Unchanged under the hood was the 3.4-liter V-6, still dishing out its 160 horsepower. New for the 5.7-liter V-8 was a change to sequential fuel injection. While this did not increase the engine's output, it improved its drivability, as well as reducing emissions and improving mileage. When the Z28 was fitted with the six-speed manual transmission, the

Left: Of course, Bright Red remained the most popular color, covering 22,740 1994 Camaros. *General Motors 2012*

Below: Arctic White proved a popular color in 1994, covering some 15,169 Camaros produced that year. *General Motors 2012*

driver had to get used to a new gizmo, CAGS, or computer-aided gear selection. Under light throttle, a solenoid would guide the shifter from first to fourth gear. Under moderate to heavy throttle, the solenoid would pull back, allowing the driver to run through all of the gears. This was intended to improve mileage and reduce emissions. It probably did those things, but it was the source of no small amount of frustration for drivers.

For drivers wanting yet more performance, and the 135 customers who were willing to kick in $310 to the bottom line, the 1LE package was still available. This race-ready road rocket was something of a sleeper on the street, as much as any Camaro could pull off the whole sleeper thing. While the 1LE didn't put any more horses under the hood, it made getting the power to the ground more efficient, as well as improving braking and overall handling. It was money very well spent.

Law enforcement had not been left out in the Camaro-less cold. Regular Production Option B4C was still on the not-for-public-consumption order form. It was still a base Camaro with an LT1 engine and other go-fast, stop-fast bits. Not inexpensive at $3,398 with a manual transmission, $3,993 with a 4L60-E automatic transmission, it was the ideal vehicle to haul down the evil speeder. Chevrolet built 668 of the crime-busters for 1994.

1994	
MODEL AVAILABILITY	two-door coupe, convertible
WHEELBASE	101.1 inches
LENGTH	193.2 inches
WIDTH	74.1 inches
HEIGHT	51.3 inches
WEIGHT	3,247 lbs
PRICE	$13,499
TRACK	60.7/60.6 inches (front/rear)
WHEELS	16 x 8 inches
TIRES	P235/55R16
CONSTRUCTION	unit body
SUSPENSION	upper and lower control arms, coil springs front/live axle, lower trailing arms, coil springs rear
STEERING	rack and pinion
BRAKES	10.9-inch disc front/9.5 x 2.0-inch drum rear
ENGINE	169-horsepower, 3.4-liter V-6; 275-horsepower, 5.7-liter V-8
BORE AND STROKE	3.62 x 3.31 inches (3.4-liter), 4.00 x 3.48 inches (5.7-liter)
COMPRESSION	9.0:1 (3.4-liter), 10.5:1 (5.7-liter)
FUEL DELIVERY	SFI (3.4-, 5.7-liter)
TRANSMISSION	five- and six-speed manual, four-speed automatic
AXLE RATIO	3.23:1 (Z28)
PRODUCTION	78,859 Base, 40,940 Z28

With 325 lb-ft of torque under its hood, the 1994 Z28 was a burnout champion. *General Motors 2012*

1995

Changes for the 1995 model year were modest, but reflected Chevrolet's habit of continually polishing a product. For the first time, RPO NW9, Acceleration Slip Regulation, also known as traction control, was available. This option could be fitted to both Z28 coupes and convertibles, and cost $450. Base Camaros powered by the V-6 were popular with buyers, and Chevrolet raised the bar with the mid year introduction of RPO L36, the 3.4-liter V-6 sourced from Buick. With 200

horsepower and 225 lb-ft of torque, it was only available with an automatic transmission, and it was priced at $350. As it was a later introduction, only 4,787 were sold, but Chevy promised that the sprightly engine would soon be available with a manual gearbox as well. Base Camaros had been saddled with a speed limiter set at 105 miles per hour due to its S-rated tires; but for 1995, the entry-level Camaro was equipped with T-rated rubber, allowing the engineers to electronically bump the top speed to 115.

The interior of the fourth-generation Camaro was a much more inviting space than the interior of the third-generation car. *General Motors 2012*

Above: The convertible version of the 1995 Camaro proved popular, with 8,024 Z28 buyers electing to go topless. *General Motors 2012*

Far right: While the looks were controversial, no one could argue with the aerodynamic efficiency of the fourth-generation Camaro. *General Motors 2012*

Right: Almost 85,000 of the 122,738 Camaros produced in 1995 had V-6 engines beneath their hoods. *General Motors 2012*

1995

MODEL AVAILABILITY	two-door coupe, convertible
WHEELBASE	101.1 inches
LENGTH	193.2 inches
WIDTH	74.1 inches
HEIGHT	51.3 inches
WEIGHT	3,251 lbs
PRICE	$14,250
TRACK	60.7/60.6 inches (front/rear)
WHEELS	16 x 8 inches
TIRES	P235/55R16
CONSTRUCTION	unit body
SUSPENSION	upper and lower control arms, coil springs front/live axle, lower trailing arms, coil springs rear
STEERING	rack and pinion
BRAKES	10.9-inch disc front/9.5 x 2.0-inch drum rear
ENGINE	160-horsepower, 3.4-liter V-6; 200-horsepower, 3.8-liter V-6; 275-horsepower, 5.7-liter V-8
BORE AND STROKE	3.62 x 3.31 inches (3.4-liter), 3.80 x 3.40 inches (3.8-liter), 4.00 x 3.48 inches (5.7-liter)
COMPRESSION	9.0:1 (3.4-liter), 9.4:1 (3.8-liter), 10.5:1 (5.7-liter)
FUEL DELIVERY	9.0:1 (3.4-liter), 9.4:1 (3.8-liter), 10.5:1 (5.7-liter)
TRANSMISSION	five- and six-speed manual, four-speed automatic
AXLE RATIO	3.42:1 (Z28)
PRODUCTION	84,379 Base, 38,359 Z28

The Z28 wasn't ignored either, with California-bound cars finally enjoying true dual exhaust. But the 5.7-liter V-8 was the only powerplant available, and its 275 horsepower was, to quote Rolls-Royce, adequate. *Road & Track* magazine found that a Z28 could cover the quarter-mile in an adequate 14.9 seconds at 93.6 miles per hour. Another magazine, *Muscle Car Review*, put a bit more whip to their Camaro and came up with a 14.1-second run with 98.07 miles per hour showing on the speedometer. Top speed was an attention-grabbing 155 miles per hour. The race-ready 1LE option attracted 106 buyers, determined to pummel the opposition, either on the track or the street. Police-ready Camaros, using RPO B4C, were still being sold, much to the disgust of speeders everywhere. The pricey option ($3,479 with a manual transmission, $4,369 with an automatic) kept all but the most well-heeled, and well-connected buyers out of the car.

When the sales figures for the model year were tallied up, Z28 sales were down slightly to 38,359, but base Camaro numbers rose a bit to 84,379, giving a total of 122,738. Unfortunately, the Ford Mustang was racking up higher sales figures, 185,986, even though the 'Stang was considerably down on power compared with the Camaro.

Above: An X-ray view of the 1996 Rally Sport. *General Motors 2012*

Right: While aerodynamically efficient, the soap-bar shape of the fourth-generation car wasn't to everyone's liking. *General Motors 2012*

1996

Big changes arrived in 1996, especially under the hood. After being off the radar for 24 years, the Camaro SS returned. This stormer wasn't exactly fresh from the factory; aftermarket company Street Legal Performance (SLP) did the upgrades. A 5.7-liter LT1 V-8 was fitted, rated at 305 horsepower, 20 higher than a stock Z28. Part of the increased output was due to the improved induction and exhaust systems. With the boost in power, the Camaro SS enjoyed improved rolling stock. Originally developed for the ZR-1 Corvette, P275-40-ZR17 tires were mounted on 17 x 9 five-spoke wheels. Yet the most dramatic change was the aggressive hood, complete with a functional NACA (National Advisory Committee for Aeronautics) scoop feeding ambient air into the induction

The scoop on the 1996 Z28 was functional. *General Motors 2012*

system. At the other end of the car, the Camaro's graceful spoiler was enhanced by a slight upturn at the trailing edge.

The production of the Camaro SS followed this path: A customer would order an SS from a dealer. The dealer request would result in a Z28 leaving the Canadian assembly line and being transported to SLP's facility in Michigan. There it would receive the "enhancements" that transformed the Z28 into a Z28/SS. Upon completion of the work, the vehicle would be returned to Chevrolet, who would then ship the Camaro SS to the ordering dealer.

All of these bits resulted in a 13.3-second quarter-mile performance, and top speed of 159 miles per hour. The SS option, RPO R7T, cost $3,999, and 2,257 were sold. When *Car and Driver* magazine evaluated a Camaro SS, the as-delivered price was $28,770, a virtual performance steal.

Upgrades weren't limited to the SS model. The Z28 got a 10-horsepower bump due to the introduction of on-board diagnostics (OBD II), which used two catalytic converters, as well as an oxygen sensor for each exhaust bank.

The racing crowd was still able to acquire the potent 1LE version, now a $1,175 option. With the beefy Z28 SS getting a lot of ink, and a lot of sales, the 1LE package really was targeted at the hardcore competitor. Only 55 were sold in 1996, but that was enough to keep the option available.

1996

MODEL AVAILABILITY	two-door coupe, convertible
WHEELBASE	101.1 inches
LENGTH	193.2 inches
WIDTH	74.1 inches
HEIGHT	51.3 inches
WEIGHT	3,306 lbs
PRICE	$14,990
TRACK	60.7/60.6 inches (front/rear)
WHEELS	16 x 8 inches
TIRES	P235/55R16
CONSTRUCTION	unit body
SUSPENSION	upper and lower control arms, coil springs front/live axle, lower trailing arms, coil springs rear
STEERING	rack and pinion
BRAKES	10.9-inch disc front/9.5 x 2.0-inch drum rear
ENGINE	200-horsepower, 3.8-liter V-6; 285-, 305-, 310-horsepower, 5.7-liter V-8
BORE AND STROKE	3.80 x 3.40 inches (3.8-liter), 4.00 x 3.48 inches (5.7-liter)
COMPRESSION	9.4:1 (3.8-liter), 10.5:1 (5.7-liter)
FUEL DELIVERY	SFI (3.8-, 5.7-liter)
TRANSMISSION	five- and six-speed manual, four-speed automatic
AXLE RATIO	3.42:1 (Z28)
PRODUCTION	34,522 Base, 8,996, RS, 17,844 Z28

As in previous years, the RS model was the base Camaro. *General Motors 2012*

Another performance option, the Special Service Package, RPO B4C, was once again offered to police departments, usually in the form of highway patrol. Sales of this pricey option ($3,369 with a six-speed manual, $4,905 with the automatic) were respectable, with 228 rolling onto highways in search of the dumb and the fast.

The SS wasn't the only news in CamaroLand. Returning after a hiatus was the Rally Sport, slotting in between the base Camaro and the Z28. Essentially it was a V-6-powered car with Z28-type lower body cladding, 16-inch wheels, and standard air conditioning. The RS was available in both the coupe and convertible models, but it wasn't as good a seller as Chevrolet had hoped, with only 11,085 units sold. Part of the reason the RS convertible racked up only 905 sales was that the base price was $22,720; the base price for a Z28 ragtop was just $24,490.

Under the hood of the base Camaro coupe and convertible, the 3.8-liter, L36 V-6 engine had set up permanent residency. This 200-horsepower mill churned out 52.6 horsepower per liter, compared with 50 ponies per liter in the LT1. Yet the Camaro, even in base trim, required that you had to pay to play. Prices had steadily climbed, and the installation of the 3.8-liter engine, while it helped to liven up the entry-level Camaro, forced the bottom line to increase $750. The list price of a base Camaro coupe, the cheapest on the menu, was $14,990, before the inevitable options sent the cost skyward. Unfortunately, prices rose across the entire Camaro line, and this took one hell of a hit on the number of units sold. At the end of the 1996 model year, sales had plummeted to just 61,362 units. That got Chevrolet's attention, but in a negative way.

There were some in the company who started muttering that the Camaro was a dinosaur whose time had come and gone. Granted, these were people you didn't really want to know, but they often controlled the purse strings. These bottom-line bottom-feeders wouldn't know a performance car if it ran over their Hewlett-Packard calculators, but they were in the shadows, clucking their tongues with every dip in Camaro sales. Fortunately, Chevrolet had a strong, pro-performance General Manager, Jim Perkins. But even his power had limits, and something had to be done to kick up sales, or else the doomsday club would have its way.

1997

It was hoped that the "something" in model year 1997 would be the Camaro acting as a Pace Car at Indianapolis again. But this time, it wouldn't be in front of a field of open-wheel missiles. Instead, the pack of cars would be NASCAR's top-shelf series competing in the Brickyard 400 held at the Indianapolis Motor Speedway in Speedway, Indiana. The three Pace Cars used at the track wore the same paint scheme that the Camaro wore at the 1969 Indy 500, white with Hugger Orange stripes. Chevrolet fit 1997 trim on the 1996 Camaros bound for actual Pace Cars duties.

To celebrate the Camaro's 30th anniversary, all 1997 Camaros wore an embroidered logo on the cloth headrests, embossed logos on leather seats. Z28s could be ordered with RPO Z4C, the 30th Anniversary Edition, costing $575. This resulted in a striking Arctic White car with Hugger Orange stripes, handsome white five-spoke wheels, and special trim. The seats could be covered in either standard white leather

with black/white cloth houndstooth inserts or optional all white leather for $499. It wasn't an ideal interior to take the kids to Baskin-Robbins. But 4,533 buyers stepped up.

Helping to commemorate the Camaro's birthday was an updated interior, which included a freshened instrument panel, more comfortable seats and a new center console. Another feature that was introduced on the 1997 was daytime running lights, in an effort to increase vehicle visibility in the name of safety.

Still in the lineup was the uber-Camaro SS, with 3,137 built. This $3,999 package above the base price of a Z28 was still built by SLP and was just as ass-kicking as before. In an interesting twist, Chevrolet had massaged their own LT1 engine into a 339-horsepower mill with a new RPO designation, LT4. SLP got a small number of vehicles with this engine, and they balanced and blueprinted each of these powerplants. Only 106 Camaro SSs were so equipped, making them even more rare and desirable.

Nothing says "Camaro" like Arctic White paint with Hugger Orange stripes. Regular production option Z4C bought the special paint, trim, and either solid white leather or white leather with black-and-white houndstooth upholstery. Embroidered seats were part of the celebratory package.

1997

MODEL AVAILABILITY	two-door coupe, convertible
WHEELBASE	101.1 inches
LENGTH	193.2 inches
WIDTH	74.1 inches
HEIGHT	51.3 inches
WEIGHT	3,306 lbs
PRICE	$16,215
TRACK	60.7/60.6 inches (front/rear)
WHEELS	16 x 8 inches
TIRES	P235/55R16
CONSTRUCTION	unit body
SUSPENSION	upper and lower control arms, coil springs front/live axle, lower trailing arms, coil springs rear
STEERING	rack and pinion
BRAKES	10.9-inch disc front/9.5 x 2.0-inch drum rear
ENGINE	200-horsepower, 3.8-liter V-6; 285-, 305-, 310-horsepower, 5.7-liter V-8
BORE AND STROKE	3.80 x 3.40 inches (3.8-liter), 4.00 x 3.48 inches (5.7-liter)
COMPRESSION	9.4:1 (3.8-liter), 10.5:1 (5.7-liter)
FUEL DELIVERY	SFI (3.8-, 5.7-liter)
TRANSMISSION	five- and six-speed manual, four-speed automatic
AXLE RATIO	3.42:1 (Z28)
PRODUCTION	29,775 Base, 9,175 RS, 21,252 Z28

The base price for all of the Camaro models took a hike upward for 1997, and this did nothing for promoting sales. A total of 60,202 Camaros were sold, compared with 108,344 Mustangs. Ford's pony car couldn't touch the Camaro in magazine comparison tests, including acceleration, handling, and braking. But the buying public put their money where their hearts were, and the Mustang had a retro vibe that the cutting-edge Camaro didn't. With the base price of a Camaro coupe coming in at $16,215, and a Z28 starting at $20,115, it didn't take a rocket scientist to realize that buyers for a pricey pony car would have to be well-heeled, such as baby boomers.

They had grown up with Mustangs and Camaros, and they wanted a modern version of the car they loved. The Ford looked like a vintage Mustang, while the Camaro looked like rolling science fiction. The buyers flocked to the Blue Oval, leaving the Camaro out in the cold. Unfortunately for fans of the Camaro, the bean-counters at Chevrolet saw this trend, and they made the decision to pull the plug on the Camaro. Its days were numbered.

With the engine in the 1997 Camaro Indy pace car replica pushed rearward, access to the back portion of the engine was a challenge. The long nose of the fourth-generation Camaro, together with the engine mounting position, gave the engineers plenty of room to fit the radiator, air conditioning evaporator, windshield washer fluid bottle, and more.

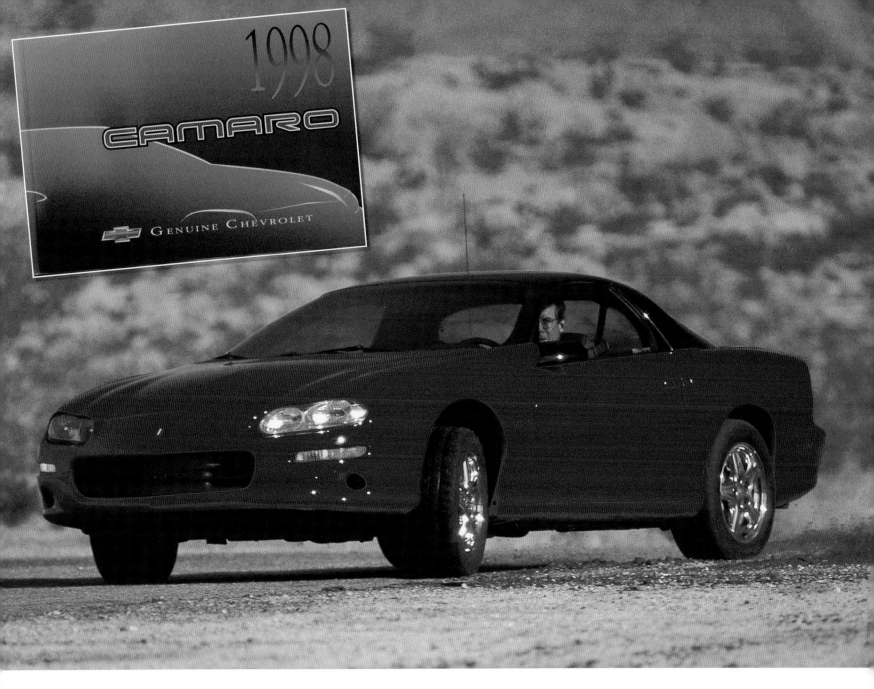

1998

Even though the clock was ticking on it, the Camaro for 1998 enjoyed some performance upgrades that helped to justify yet another round of price increases. A base Camaro Coupe started at $17,150, and the base price of a Z28 convertible was $27,975. The famed aluminum LS1 Gen III engine from the Corvette, rated at 305 horsepower, replaced the LT1. This engine in the Camaro was very conservatively rated. It was an unwritten rule at Chevrolet that no other vehicle in the roster would have more power than the Corvette. So on paper, the 'Vette was stronger; but on the road, a heavy foot told a different story. Buyers of the Z28 could choose between the automatic or manual transmission at no cost. This hot Z28 was capable of stunning performance; *Car and*

Driver magazine flogged one down the drag strip in 13.8 seconds, bone stock. The Camaros in the 1960s couldn't deliver that number.

The SS package was now RPO WU8 and was somewhat in-house. A large amount of SS goodies that used to be put on a Z28 were now installed at the factory. However, Camaro SSs were still shipped to SLP for installation of the composite hood, spoiler, and other parts. Horsepower in the Camaro SS was 320, the LS1 sucking in air through the functional cold-air hood scoop. Because of the partial buildup at the factory, the price of the SS option actually decreased, to $3,500. This didn't hurt SS sales at all, as 3,025 were sold. In a model year when the total number of Camaros built was 54,026, the sales of that many SS models represented a considerable

For model year 1998, Z28s used the Corvette's LS1 5.7-liter V-8 engine, rated at 306 horsepower. In order to enjoy fun like this, buyers had to spend $20,470 for the coupe or $27,450 for the convertible. Either one could significantly shorten the life of the rear tires with ease.

1998

MODEL AVAILABILITY	two-door coupe, convertible
WHEELBASE	101.1 inches
LENGTH	193.2 inches
WIDTH	74.1 inches
HEIGHT	51.3 inches
WEIGHT	3,331 lbs
PRICE	$17,150
TRACK	60.7/60.6 inches (front/rear)
WHEELS	16 x 8 inches
TIRES	P235/55R16
CONSTRUCTION	unit body
SUSPENSION	upper and lower control arms, coil springs front/live axle, lower trailing arms, coil springs rear
STEERING	rack and pinion
BRAKES	10.9-inch disc front/9.5 x 2.0-inch drum rear
ENGINE	200-horsepower, 3.8-liter V-6; 305-, 315-, 320-horsepower, 5.7-liter V-8
BORE AND STROKE	3.80 x 3.40 inches (3.8-liter), 3.90 x 3.62 inches (5.7-liter)
COMPRESSION	9.4:1 (3.8-liter), 10.1:1 (5.7-liter)
FUEL DELIVERY	SFI (3.8-, 5.7-liter)
TRANSMISSION	five- and six-speed manual, four-speed automatic
AXLE RATIO	3.42:1 (Z28)
PRODUCTION	33,973 Base, 20,053 Z28

percentage of the model run. However, the Mustang moved 170,642 units, outselling the Camaro three-to-one.

But changes to the Camaro weren't limited to beneath the huge hood. If fact, the hood was a touch taller to accommodate the LS1 engine. The most noticeable difference from the 1997 Camaro was the new nose, with its flush halogen headlights and huge simulated grill. For years, all engine cooling air entered the engine compartment from beneath the nose, and that didn't change for 1998. The gaping maw was strictly a design element that did not ingest any air.

The Rally Sport, which had reappeared in 1997, was yanked again, replaced with RPO Y3F, essentially a $1,755 body cladding kit with an upturned tail spoiler. This option cost $1,480 on the Z28, and a heady 7,411 people sprang for it.

Below: The back seat of a 1998 Camaro Z28 was an ideal place to sit—if you were seven years old. Fit for kids and people you didn't like, the +2 part of the seating arrangement did mean that the Camaro was a four-seater, giving it lower insurance premiums.

1999

Not much changed for model year 1999. Oh, the V-6 engine in the coupe and convertible now utilized electronic throttle control and traction control, and three new colors were introduced. By switching to a non-metallic (plastic) gas tank, the capacity was increased from 15.5 gallons to 16.8. Chevrolet was loathe to put any real money in a platform that was nearing its expiration date.

This generation of Camaro had been around for quite some time and, frankly, General Motors didn't have the resources to create a fresh-sheet Camaro. As Larry Edsall notes in *Camaro, A Legend Reborn*, "[T]he car's chassis was based on old technology, and new federal safety regulations were coming, regulations that would demand a major investment to update the car's architecture. It was an investment General Motors was neither in a position nor willing to make. There

Above: Sunrise with a clear horizon, a 1999 Camaro Z28 hurtles toward the rising sun. The rear spoiler was developed as a subtle extension of the shoulder line, allowing air to flow down the rear window onto the spoiler without breaking the vehicle's visual continuity. With the large rear window, over-the-shoulder visibility was rather good.

The RS convertible proved almost as popular as the Z28 drop top. *General Motors 2012*

Above: Unlike the third-generation Camaro's interior, which was very linear and crisply styled, the fourth-generation F-body's cockpit was filled with organic curves and soft-touch materials. But make no mistake; the rear seats were seats in name only, unless your day job was as a jockey.

Top: A plastic intake manifold saved weight under the hood of the 2000 Camaro SS, an important consideration for engineers to consider when developing an engine. The insulated fuel line for the fuel injection system is visible in the foreground. The heat-resistant wrapping helps to keep the fuel cool before it reaches the fuel injectors.

Above: Chromed, five-spoke wheels were part of the Z28 package in 1999. Bright Red, paint code 81, was one of the most popular Camaro colors that year, with 7,788 at the receiving end of the spray gun. The most popular color that year was black, with 8,742 units looking like night on wheels.

Left middle: The loooong front overhang of the fourth-generation Camaro is evident on this 1999 Z28. Designed to cleave the air smoothly while maintaining stability at high speeds, care had to be taken on steep driveways to avoid gouging the road, to say nothing about damaging the underside of the car.

Left: The 1999 Camaro Z28 continued to use a tri-color taillight assembly, neatly packaged under the swooping spoiler. Unlike the bright exterior trim used during the Camaro's early days, the fourth-generation models eschewed chrome for black finished bits to highlight design elements.

Above: Standard with every 1999 Camaro Z28 was blurred pavement—and lots of it. Since its inception, the Z28 has been engineered to handle curves at least as well the straight-line duties. With the advent of radial tires, Chevrolets engineers continually upgraded the ability of the Z28 to generate significant lateral g's, in this case, 0.81.

Below: The front seats in the 1999 Camaro Z28 were low, almost on the vehicle's floor. They enjoyed a 6.0-inch range of fore/aft motion, and the well-bolstered seats held the driver and front passenger in place during spirited maneuvers. People in the back, meanwhile, were chewing on their knees and looking forward to escaping.

Above: The 5.7-liter V-8 engine in the 1999 Camaro Z28 used a 10.1:1 compression ratio to help generate its 305 horsepower. Premium fuel was required, but not as much as you would think; the EPA rated the Camaro's fuel economy at 18 mpg in the city, 27 mpg on the highway. A very tall top gear, and a 3.42:1 rear axle ratio, made high-speed cruising a low-stress event.

1999

MODEL AVAILABILITY	two-door coupe, convertible
WHEELBASE	101.1 inches
LENGTH	193.5 inches
WIDTH	74.1 inches
HEIGHT	51.3 inches
WEIGHT	3,331 lbs
PRICE	$17,240
TRACK	60.7/60.6 inches (front/rear)
WHEELS	16 x 8 inches
TIRES	P235/55R16
CONSTRUCTION	unit body
SUSPENSION	upper and lower control arms, coil springs front/live axle, lower trailing arms, coil springs rear
STEERING	rack and pinion
BRAKES	10.9-inch disc front/9.5 x 2.0-inch drum rear
ENGINE	200-horsepower, 3.8-liter V-6; 305-, 320-horsepower, 5.7-liter V-8
BORE AND STROKE	3.80 x 3.40 inches (3.8-liter), 3.90 x 3.62 inches (5.7-liter)
COMPRESSION	9.4:1 (3.8-liter), 10.1:1 (5.7-liter)
FUEL DELIVERY	SFI (3.8-, 5.7-liter)
TRANSMISSION	five- and six-speed manual, four-speed automatic
AXLE RATIO	3.42:1 (SS)
PRODUCTION	24,901 Base, 17,197 Z28

were too many other priorities, both for car buyers and automakers. As General Motors Vice Chairman and car guru Bob Lutz would later explain, the F-body platform beneath the Camaro and Firebird had become 'a bridge too far in bad packaging.'" And while changes don't in themselves attract hordes of customers, another price increase did little to pull customers into dealerships.

With the base price of a Camaro convertible coming in at $22,740, and a Z28 coupe selling for $21,485, there was ambulance-grade sticker shock. This was painfully evident at the end of the year, when total Camaro sales were only 42,098 units. Once again, the Mustang outsold the Camaro by a three-to-one margin, with 133,637 'Stangs being titled. Ironically, sales of the most expensive Camaro model, the Z28 SS, were higher than ever, with 4,829 being built. People who knew real performance knew that the Z28 SS was one hell of a performance bargain, probably the best in America. But the rest of the Camaro languished.

2000

Carry-over was the name of the game in 2000. New aluminum wheel designs were introduced, new colors and fabrics were available in the interior, and radio controls on the steering wheel was an option. The sales numbers were stagnant, leaning toward down. With 45,461 total Camaros sold, the champagne wasn't exactly flowing at Chevrolet. Yet 17,383 customers were willing to plunk down $21,800 for a Z28 coupe, and another 3,036 Z28 ragtops listing for $28,900 were removed from dealer lots. The famous Hurst shifter was available from the factory, costing $325.

Racing drivers lamented that the 1LE option was gone; only 74 had been sold in 1999. There were no changes under the hood, which isn't exactly a bad thing when the Z28 could sprint the length of a drag strip in 13.8 seconds, and a Camaro SS could follow that up with a 13.5-second run at 105.7 miles per hour. The Camaro SS would top out at 162 miles per hour. If it's not broken, don't fix it.

Left: An air bag filled the center section of the 1999 Camaro Z28's two-spoke steering wheel. The speedometer was not intended to deceive. With a slightly detuned Corvette 5.7-liter engine making 305 horsepower, the Z28 could force the speedo needle to swing virtually through a full arc.

Above: This vantage point shows the subtle curves that Chevrolet styled into the 2000 Camaro SS. The rear spoiler carried its arc around the entire base of the rear window. The long-hood/short-deck proportions of a classic pony car have never been more evident than from this perspective.

Far left: Another name from the Camaro's past, Hurst was tapped to provide the six-speed manual transmission's shifter in the 2000 Camaro SS. A very beefy unit, it took a firm hand to grab a gear, but once in, there was no doubt that you'd gotten another cog. It had the solidity of a vault door.

Left: When the Super Sport label landed on the Camaro in 1967, it denoted a bad-to-the-bone muscular vehicle that didn't take any prisoners. It was no different on the 2000 Camaro SS. The performance package wasn't cheap at $3,950, but as in years past, the Super Sport package was value rich.

The raw performance of the 2000 Camaro SS was knocking on Big Brother Corvette's door, courtesy of a stellar LS1 5.7-liter engine that used 3.5-inch stainless-steel exhaust to improve breathing and generated a tire-melting 320 horsepower. Coupled with 345 lb-ft of torque, the Goodyear Eagle F1 tires didn't have much of a future. In the center of the hood of the 2000 Camaro SS is a functional hood scoop, drawing cool outside air into the engine compartment. This feature helped *Road & Track* get the car down the quarter-mile in 13.9 seconds at 105.7 miles per hour. Only 3,017 Camaro SS's were built that year.

Above: Hammering down a drag strip in 13.9 seconds at 105.7 miles per hour would, in 1969, qualify the 2000 Camaro SS as a competitor to the ZL-1, but with the increasing use of computers in automobiles, the newer car was just as comfortable going to the grocery store as collecting timing slips.

2000

MODEL AVAILABILITY	two-door coupe, convertible
WHEELBASE	101.1 inches
LENGTH	193.5 inches
WIDTH	74.1 inches
HEIGHT	51.3 inches
WEIGHT	3,360 lbs
PRICE	$17,375
TRACK	60.7/60.6 inches (front/rear)
WHEELS	16 x 8 inches
TIRES	P235/55R16
CONSTRUCTION	unit body
SUSPENSION	upper and lower control arms, coil springs front/live axle, lower trailing arms, coil springs rear
STEERING	rack and pinion
BRAKES	10.9-inch disc front/9.5 x 2.0-inch drum rear
ENGINE	200-, 205-horsepower, 3.8-liter V-6; 305-, 320-horsepower, 5.7-liter V-8
BORE AND STROKE	3.80 x 3.40 inches (3.8-liter), 3.90 x 3.62 inches (5.7-liter)
COMPRESSION	9.4:1 (3.8-liter), 10.1:1 (5.7-liter)
FUEL DELIVERY	SFI (3.8-, 5.7-liter)
TRANSMISSION	five-, six- speed manual, four-speed automatic
AXLE RATIO	3.42:1 (SS)
PRODUCTION	25,042 Base, 20,419 Z28

Above: This is where 320 horses live. The slightly de-tuned 5.7-liter Corvette engine used a center-mount air cleaner to maximize ambient airflow into the intake manifold. Note the significantly large area between the radiator and the front edge of the bumper housing. This throttle-by-wire engine was set far back in the chassis.

Right: A brute in svelte clothing, the 2000 Camaro SS could accelerate to 60 miles per hour in just 5.5 seconds and dash the length of a drag strip in 13.9 seconds. Having a 5.7-liter, 320-horsepower V-8 under the long hood put the life expectancy of the Goodyear F1 tires into doubt. The handsome wheels reduced unsprung weight as well as looked good.

2001

The following year would have the lowest production numbers in Camaro history. With only 29,009 units built, the 2001 Camaro production run was actually shortened to allow for more 2002s to be built. By this time, the public was very aware that the Camaro would be a memory after 2002, and Chevrolet figured that a lot of people would wait for a chance to get a last-year version. This proved correct. But this isn't to say that the 2001 Camaro was left alone; the Z28 benefited from some engine work. Both the regular Z28 and the SS got a new intake manifold, sourced from the LS6 engine. That powerplant was used in the fifth-generation Corvette Z06. Its superior flow rate, along with a revised camshaft and pulling of the EGR system boosted the rated power of the Z28 to 315 horsepower, and 325 for the SS model. The SS version also received a power steering cooler. Manual transmission-equipped V-8 Camaros utilized the hardy LS6 clutch.

The heart of the SS package was an aluminum 346-cubic-inch LS1 V-8 engine that pumped out 325 horsepower. *General Motors 2012*

Above: Chevrolet sold just 1,452 Z28 convertibles in 2001. *General Motors 2012*

Right: While the 2001 Camaro was the top-performing car of its day, sales were dismal; Chevrolet sold just 29,009 units, effectively killing the car's future. *General Motors 2012*

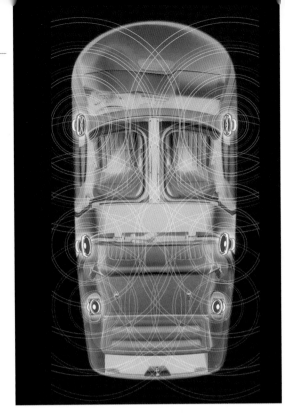

Above: What Superman sees when he looks at a Camaro.
General Motors 2012

Below: Chevrolet equipped 6,332 Camaros with the WU8
SS package in 2001. *General Motors 2012*

2001

MODEL AVAILABILITY	two-door coupe, convertible
WHEELBASE	101.1 inches
LENGTH	193.5 inches
WIDTH	74.1 inches
HEIGHT	51.2 inches
WEIGHT	3,331 lbs
PRICE	$17,650
TRACK	60.7/60.6 inches (front/rear)
WHEELS	16 x 8 inches
TIRES	P235/55R16
CONSTRUCTION	unit body
SUSPENSION	upper and lower control arms, coil springs front/live axle, lower trailing arms, coil springs rear
STEERING	rack and pinion
BRAKES	10.9-inch disc front/9.5 x 2.0-inch drum rear
ENGINE	200-, 205-horsepower, 3.8-liter V-6; 310-, 325-horsepower, 5.7-liter V-8
BORE AND STROKE	3.80 x 3.40 inches (3.8-liter), 3.90 x 3.62 inches (5.7-liter)
COMPRESSION	9.4:1 (3.8-liter), 10.1:1 (5.7-liter)
FUEL DELIVERY	SFI (3.8-, 5.7-liter)
TRANSMISSION	five-, six-speed manual, four-speed automatic
AXLE RATIO	3.42:1 (SS)
PRODUCTION	16,357 Base, 12,652 Z28

2002

For its last appearance on the stage, the 2002 Camaro was little changed from the previous year. Buyers of SS models, both coupes and convertibles, could pony up an additional $2,500 for RPO Z4C, the 35th Anniversary Edition. This resulted in a Bright Red exterior, with an ebony leather interior with gray inserts. The headrests were the site of another meeting with the embroiderer, as anniversary logos were sewn in.

Fancy silver exterior body stripes that turned into a flowing checkered flag pattern were part of the anniversary option. Camaro SS coupes that were fitted with the 35th Anniversary Edition package had to have T-tops. The commemorative package was popular, with 3,369 sold. They were all powered by the LS1 5.7-liter engine, rated at 325 horsepower. Camaro SS's with a manual transmission were fitted with 3.42:1 rear axle gears, while automatic-equipped vehicles had a set of 3.23:1 gears in the pumpkin. The first 45 35th Anniversary Edition convertibles were used as festival cars at the Brickyard 400 race at the Indianapolis Motor Speedway.

The V-6-equipped Camaro soldiered on for 2002, as the 3.8-liter mill was still generating 200 horsepower; and RPO Y87, the Performance Handling Package, was still on the order form. This package retailed for $275 and included a Torsen limited-slip differential, quicker (14.4:1) steering, and four-wheel disc brakes. Good value.

Prices had risen again for 2002. A base Camaro coupe listed for $18,455, while the ragtop version of the entry level F-body went for $26,450. Z28 buyers needed deeper pockets, as the coupe version stickered at $22,870, while the convertible was going for a lofty $29,965. Serious money; but then, the Z28 was a serious car.

Over the years, a number of mandatory options had been listed; and in 2002, there was a real peach: RPO R6M. This was the New Jersey Surcharge for "warranty enhancement," a $93 mandatory option charged to 789 buyers. This little gift was also given to Corvette buyers. New Jersey fattened their state coffers with schemes to extract more money from taxpayers, and new vehicle buyers weren't spared.

With the demise of the Camaro looming, Chevrolet used natural attrition to thin the ranks of workers at the factory, but the thinning process was a bit more effective than Chevy had planned. When the 2002 Camaros were rolling down the line, there weren't enough employees to build enough cars to fill all of the orders. But the assembly line had to shut down by September 1, 2002, as after that date the Camaro would no longer meet federal head-impact standards. So on August 27, 2002, the last fourth-generation Camaro, a red Z28 convertible, rolled off the line and into history. A total of 41,776 2002 Camaros had been built. The following day, the Ste. Therese assembly line shut down. For all intents and purposes, the Camaro was gone. But, as the saying goes, never say never. Once again, the Mustang would bring the Camaro to life.

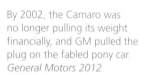

By 2002, the Camaro was no longer pulling its weight financially, and GM pulled the plug on the fabled pony car. *General Motors 2012*

In its last year of production, the Camaro reversed a long-standing tradition, and the Z28 version outsold the base coupe. *General Motors 2012*

After the 2002 model year, the Camaro appeared to have motored off into the sunset for good. *General Motors 2012*

Inset: The last of the original Camaros was also the best, inside and out. But it wasn't enough. *General Motors 2012*

This 2001 Camaro SS is doing its part for global warming as it heats up its rear tires prior to launching down the drag strip. The nearly flush grille and headlight clusters improved the aerodynamics of the F-body, improving fuel economy and giving the car a sinister look.

MODEL AVAILABILITY	two-door coupe, convertible
WHEELBASE	101.1 inches
LENGTH	193.2 inches
WIDTH	74.1 inches
HEIGHT	51.3 inches
WEIGHT	3,310 lbs
PRICE	$18,455
TRACK	60.7/60.6 inches (front/rear)
WHEELS	16 x 8 inches
TIRES	P235/55R16
CONSTRUCTION	unit body
SUSPENSION	upper and lower control arms, coil springs front/live axle, lower trailing arms, coil springs rear
STEERING	rack and pinion
BRAKES	10.9-inch disc front/9.5 x 2.0-inch drum rear
ENGINE	200-, 205-horsepower, 3.8-liter V-6; 310-, 325-, 345-horsepower, 5.7-liter V-8
BORE AND STROKE	3.80 x 3.40 inches (3.8-liter), 3.90 x 3.62 inches (5.7-liter)
COMPRESSION	9.4:1 (3.8-liter), 10.1:1 (5.7-liter)
FUEL DELIVERY	SFI (3.8-, 5.7-liter)
TRANSMISSION	five-, six-speed manual, four-speed automatic
AXLE RATIO	3.42:1 (SS)
PRODUCTION	16,971 Base, 24,805 Z28

35 years
1967-2002

5

GENERATION FIVE

2010-Today Resurrection

Above: Halo lights surround the high-intensity headlights recessed within the grille of a 2010 Camaro SS. With a vehicle as capable of serious speed on the road, it's important to be able to see ahead at night, and the HID lighting gives a driver a measure of confidence.

Opposite: Since the first Camaro rolled out in late 1966, this is what Chevrolet's pony car has been good at; blurring the pavement. The 2010 Synergy Green Camaro, with its healthy V-6 engine, has less weight over the front wheels than the V-8-packing SS model, and handling is the better for it.

When the Camaro went out of production in late 2002, the pony car market consisted of a single make. Ford found their Mustang playing in a field of one. This translated to good sales numbers, with 150,895 sold in the 2003 model year. These were quantities that made Chevrolet green with envy. At the same time, Ford started development of a new Mustang, one that would summon the spirit of the Mustangs from the 1960s. In late 2004, the new 2005 Mustang debuted to acclaim. Sales of the Mustang rose from 129,858 units in 2004, to 160,975 in 2005. The 2005 Mustang was a perfect blend of retro styling with a modern spin. The press loved it, the public loved it; only Chevrolet hated it.

General Motors realized that the market for a performance car with devilish good looks was alive and well. But the company didn't have anything in America that could be tossed into the ring with the Mustang. Looking around the globe, the Holden Monaro from Australia was seen as a possible contender.

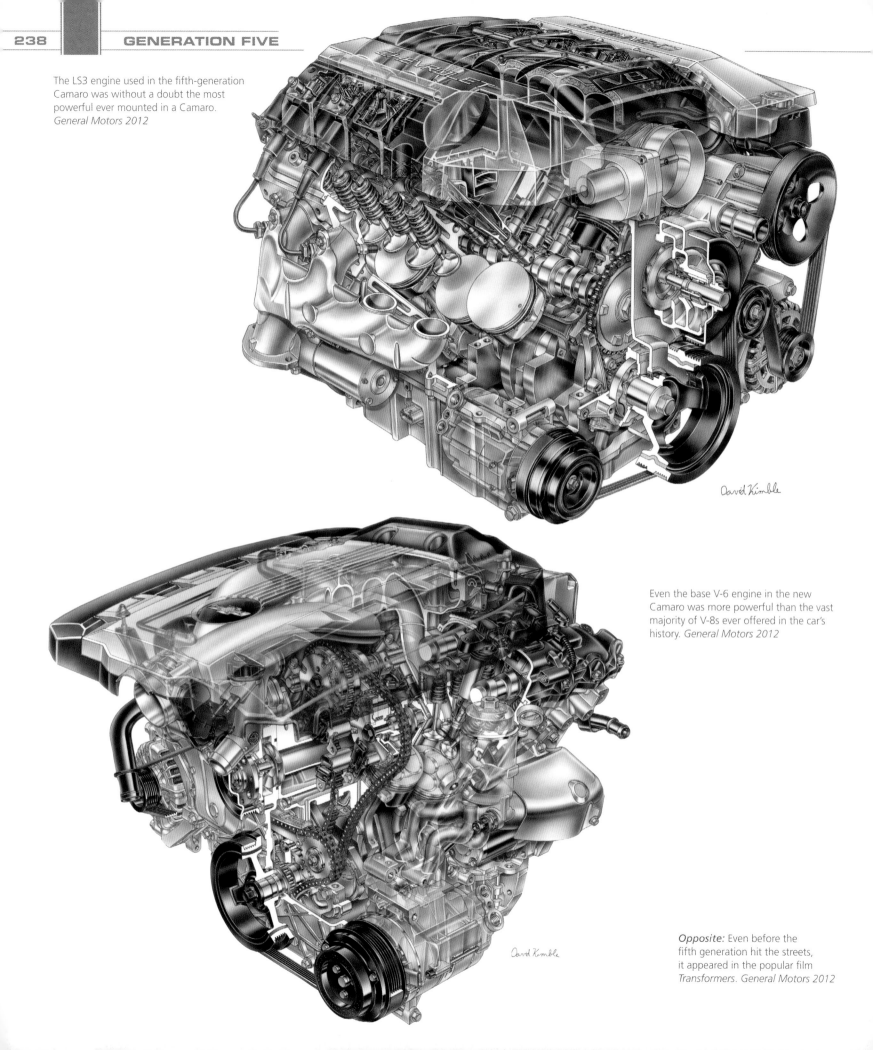

The LS3 engine used in the fifth-generation Camaro was without a doubt the most powerful ever mounted in a Camaro. *General Motors 2012*

Even the base V-6 engine in the new Camaro was more powerful than the vast majority of V-8s ever offered in the car's history. *General Motors 2012*

Opposite: Even before the fifth generation hit the streets, it appeared in the popular film *Transformers. General Motors 2012*

It was a very capable road car, created in the mold of the original muscle car. Unlike the original muscle cars, the Monaro enjoyed a fully independent suspension, giving it handling older muscle cars could only dream of. The program to create a Camaro from a Monaro was called GMX280. But this 5.7-liter front-engine, rear-drive platform was a right-hand drive machine that would need a lot of expensive massaging to get it ready for North American roads. Eventually, the GMX280 program was axed, but it led General Motors to consider another storied nameplate to bolt onto the car. With a limited budget, most of the money was spent in converting the platform to a left-hand-drive configuration. That left little coin to spice up the Monaro's bland looks. GM had, in North American chairman Robert Lutz, a real car guy making decisions; and even though the Monaro had all the visual verve of a melted lozenge, it was a formidable street bruiser. It could sprint to 60 miles per hour in just 5.5 seconds and cover the quarter-mile in 14 seconds. So General Motors christened it a Pontiac GTO and tossed it into the deep end.

Above: The 2010 Camaro SS rides on P245/45R20 summer tires surrounding 20 x 8 aluminum wheels in the front, while 20 x 9 alloy wheels keep the back of the car off the ground. The big rubber helps the Camaro SS generate 0.90g on a skid pad.

Above: In an effort to gauge public reaction to the Camaro concept vehicle, Chevrolet asked comedian and car guy Jay Leno to drive the car to the famous Friday night cruise night at Bob's Big Boy restaurant in Toluca Lake, California. The reaction was, well, enthusiastic.

Right: Chevrolet put SS badging in conspicuous locations around the 2010 Camaro SS, including the front seat headrests. Sewn into the leather upholstery, the performance logo has been a symbol of automotive excitement for over 50 years.

The results were a foregone conclusion. It was a sales disaster. Fewer than 14,000 were sold during the entire production run. American muscle car enthusiasts want their muscle cars to look like they have muscles. The 2004 GTO looked like a well-used bar of soap. General Motors performance enthusiasts who wouldn't be caught dead in a Ford product stayed away from the GTO in droves and continued to put miles on their old rides. The GTO only lasted three years, as the 2006 model was the last.

In the shadows, things were about to change at GM in general, and Chevrolet in particular. Ed Welburn, GM vice president of global design, was having a beer with recently hired head of Chevrolet's Advanced Design Studio, Bob Boniface. As Darwin Holmstrom recounts in his book *Camaro: Forty Years*, "Like Boniface, Welburn was a huge Camaro fan, but he know that such a project [building a new Camaro] would meet resistance from GM management. 'Just don't tell anybody,' Welburn advised. Boniface began working on a concept in April 2005. About halfway through the year, designer Tom Peters began working on the Camaro project. Peters had been the project leader for the sixth-generation Corvette. When the Australian Division redesigned the Zeta architecture for the next-generation Monaro, it engineered the car to be either right-hand of left-hand drive, which meant that developing versions for the North American market was a much-less-expensive proposition."

Peters and his crew took significant styling elements from past Camaros, especially the iconic 1969 model. Their work was unveiled at the annual North American International Auto Show in Detroit, Michigan, on January 6, 2006, as their Camaro Concept was driven onto the stage by Bob Lutz. The Camaro Concept was penned by Sangyup Lee, and it rode on a 110-inch wheelbase, longer than its predecessor. But the overall vehicle length was 186.2 inches, 7 inches

The 2010 Synergy Green Camaro debuted at the 2010 Detroit International Auto Show, and it's based on the 1LT model, with a 304 horsepower, 3.6-liter direct-injection V-6. This engine uses an 11.3:1 compression ratio to churn out 273 lb-ft of torque, enough to vault to 60 in 5.9 seconds.

The svelte shape of the 2010 Camaro SS can tear down the quarter-mile in just 13.0 seconds at 111 miles per hour. Zero to 60 requires 4.7 seconds. The vent at the leading edge of the hood is a Super Sport exclusive; the V-6 Camaro hood is a smooth affair.

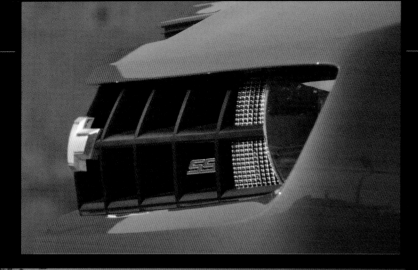

Top: More than five thousand welds go into the making of a 2010 Camaro SS, giving the vehicle a structural rigidity never seen before in a Camaro. In Super Sport configuration, the aggressive nose of the vehicle can cleave the air at an electronically limited 157 miles per hour.

Above: Deeply inset primary gauges in the instrument panel are reminiscent of the panel treatment in a 1969 Camaro, as well as the secondary instruments mounted in front of the shifter. Heavily bolstered seats help keep the driver behind the wheel while busy carving canyons.

Right: If a 2010 Camaro SS is equipped with the six-speed automatic transmission, a set of shifter paddles are mounted on the back side of the steering wheel. The paddle on the right side of the wheel is marked with "+" and controls upshifts, while the left-side paddle in labeled "-" and handles downshifts.

Top: The 6.2-liter (376-cubic-inch) V-8 redlines at 6,000 rpm and is rated at two horsepower levels depending on which transmission is used. Camaro SS's equipped with a Hydra-Matic 6L50 six-speed automatic come in at 400 horses, while TR6060 six-speed manual users enjoy 426 ponies.

Above middle: While these secondary gauges in the 2010 Camaro SS are handy, they aren't exactly placed for quick reading. But with 400 horsepower in this automatic-transmission-equipped example, the driver's attention is most likely directed out the narrow window across the long hood.

Above: Chevrolet teamed up with famed Italian brake maker Brembo to ensure the 2011 Camaro SS could stop as well as it could go. Within the 20-inch aluminum wheels are 14.0-inch rotors in the front, while huge 14.4-inch rotors fill the rear wheels. Four-piston aluminum calipers clamp down.

It's a small touch, but the SS emblem on the rim of this 2010 Camaro SS is a subtle, tasteful reminder to the driver that they're not behind the wheel of a blandmobile. The leather-wrapped steering wheel, with contrasting stitching, is a beefy, gripable affair, communicating road feel to the driver's hands.

Opposite: For 2010, the Camaro Synergy was available in this subtle hue, cleverly named Synergy Green. Like all Camaros that preceded it, the 2010 version retains the long-hood/short-deck formula that is key to a performance muscle car. In a vehicle as eye-popping as a Synergy Green 2010 Camaro, it is tough to keep your foot out of the engine. This modern muscle car will cover the quarter-mile in 14.5 seconds at 99 miles per hour. It's electronically limited to 157 miles per hour, fast enough to make a driver's license disappear forever.

shorter than the 2002 Camaro. The 3,750-pound automobile had a width of 75.5 inches, 1.4 inches wider than a 2002 model, while the height of the new car came in at 54.2 inches; the fourth-generation coupe was 51.3 inches tall.

When the Camaro Concept rolled out into the glare of the lights, the assembled press was told that the production car would become available in 2008. On August 10, 2006, the then current GM Chairman and CEO Rick Wagoner confirmed what everyone suspected: a production version of the Camaro Concept car was being green-lighted. He stated that Chevrolet intended to start production at the end of 2008, and the vehicle would hit showrooms early in 2009, with the vehicle badged as a 2009 model.

Another version, the Camaro Convertible Concept, was shown to the assembled press at the 2007 North American International Auto Show on January 6, 2007. While the concept car looked like a coupe just shorn of its top, it had significant structural differences from the closed car. The rear fenders and trunk were different from the coupe. Underlining reinforcement ensured structural rigidity, as nobody wants a FlexiFlyer on the road.

Production of the fifth-generation Camaro would be at GM's Car Assembly at Oshawa, Canada, a facility that originally opened in 1953. Chevrolet revealed that two models of base Camaro would be available, starting at $22,680. Both LS and LT trim versions would use a 312-horsepower 3.6-liter DOHC V-6. This engine would get the Camaro down the quarter-mile in 14.5-seconds at 99 miles per hour. Buyers wanting to replace their rear tires with regularity tended to choose the $31,795 SS package.

Two different engines were used, both displacing 6.2 liters. For Camaro SS's equipped with a Tremec TR6060 six-speed manual transmission, the buyer would find RPO LS3, generating 426 horsepower at a lofty 5900 rpm, and 420 lb-ft of torque at 4600 revs. For customers taking home a six-speed Hydra-Matic 6180 automatic transmission, their Camaro SS would be fitted with RPO L99, its 6.2 liters pumping out 400 horsepower at 5900 rpm, and 410 lb-ft of twist at 4300 rpm. The editors at *Car and Driver* hurled one of these down a drag strip, coming in with a 13.0-second 111-miles per hour run. The L99 engine employed an active fuel management system that deactivated four cylinders during light-load conditions, thus improving fuel economy. Both V-6- and V-8-powered Camaros were electronically limited to 157 miles per hour.

For the first time in the Camaro's history, it was equipped with a fully independent rear suspension that used a 4.5-link setup, progressive-rate coil springs, stabilizer bar, and fully adjustable camber and toe. Disc brakes on every corner were standard on all Camaros, and SS models used race-bred Brembo calipers clamping 14-inch rotors on the front, 14.4-inch in the rear. Surrounding the brakes on the Camaro SS were huge 20-inch aluminum wheels, wrapped in Pirelli PZero rubber. Available on the LT and SS models was an RS appearance package, which included trick headlight halo rings, a rear spoiler, and RS-only wheels and taillights.

The first production Camaro rolled down the line on March 16, 2009, and it was a 2010 SS model, painted Silver Ice Metallic with Cyber Grey Metallic stripes. This vehicle was auctioned at the 2009 Barrett-Jackson Collector Car Auction on January 17, 2009, selling for $350,000. All of the proceeds went to the American Heart Association.

Being the first year for a new generation Camaro, it was inevitable that an Indy 500 Pace Car would be created. The actual Pace Car was painted Inferno Orange; and it was an SS model with the RS package, equipped with an automatic transmission. Chevrolet built 295 commemorative editions to sell to the public, with an MSRP of $41,950. Standard fifth-generation Camaro sales exceeded expectations, as total 2009 calendar sales were 61,648—outstanding, considering the vehicle didn't hit the showroom until April 2009.

Hot on the heels of the release of the newest Camaro was the release of a blockbuster movie, *Transformers*, in which a new Camaro, finished in a yellow and black paint scheme,

Interior panels painted Synergy Green were part of the Synergy package. A thin ribbon of light is built into the upper portion of the door panel trim, bathing the interior with a soft glow at night. The three-spoke steering wheel is adjustable for height and reach.

Right: A small rear spoiler is part of the Synergy Green package for 2010, and the dual exhaust tips are as functional as they are attractive. With a resonator in each exhaust stream, the V-6-equipped Camaro delivers an aggressive sound without tripping nearby car alarms.

were festooned with logos from the movie, as well as a bold black hood stripe. Sales of the option exceeded estimates; Chevrolet hoped to move 1,500 units, but eventually 1,784 were sold.

Another limited-run Camaro that was unveiled for 2010 was the Synergy Special Edition, a street-legal version of the show car seen at the 2009 SEMA show. Using RPO GHS, it came with a retina-burning shade of Synergy Green with Cyber Gray Rally Stripes. Also fitted to the car was the rear spoiler from the SS. The interior was a colorful mixture of green and black. This low-profile V-6-powered coupe listed for $26,790, and Chevrolet built 2425 units in a three-month period.

Due to the success of the Synergy Green package in 2010, Chevrolet expanded the Synergy Series for 2011, with four colors offered: black, white, Victory Red, and Cyber Gray. Available on the 2LT and 2SS models (a V-6 and V-8, respectively) it had special graphics and trim, body mods, and interior bits unique to the Synergy Series, including red stitching on gray leather-appointed seats, steering wheel, shifter boot, center console lid, shift knob, and door armrest.

played a prominent part. Ironically, the actual vehicles in the movie were modified Pontiac GTOs wearing panels made from molds pulled from the Camaro Concept car. The movie cars were built by Mustang wrangler Steve Saleen. At the San Diego Comic-Con on July 22, 2009, Chevrolet revealed that a "Transformer Special Edition" would be available. Coded RPO CTH, the $995 appearance package could only be used on Camaros ordered in Rally Yellow, and the finished vehicles

Above: Beneath this vast expanse of black plastic lives a potent V-6 all-aluminum engine, rated at 304 horsepower at 6400 rpm, and 273 lb-ft of torque at 5200 revs. The redline on this powerplant is a lofty 7000 rpm, and the tuned exhaust gives it a suitably aggressive note.

Left: The Synergy Green Camaro package debuted at the Specialty Equipment Manufacturers Association (SEMA) show in 2009, and for a base price of $27,000, it delivered a lot of bang for the buck. Echoing the trend in automobiles, the package includes full conductivity and the ability to use a wide range of entertainment platforms.

Above: Chevrolet introduced its 2011 Camaro Convertible at the 2010 Los Angeles International Auto Show. The automaker originally intended to release the convertible at the same time as the coupe, but Chevy's engineers didn't feel that the ragtop was ready at that time. The release date was pushed back until they had a car they were happy with.

Right: With the folded convertible top hidden beneath a cover, the 2011 Camaro convertible enjoys superb visibility. Here, the press crowds around the vehicle at its debut at the Los Angeles International Auto Show in late 2010.

With its advanced suspension and potent engine, the new Camaro was destined for the racetrack. *General Motors 2012*

2010

MODEL AVAILABILITY	two-door coupe
WHEELBASE	112.3 inches
LENGTH	190.4 inches
WIDTH	75.5 inches
HEIGHT	54.2 inches
WEIGHT	3750 lbs
PRICE	$31,040 (SS)
TRACK	63.7/63.7 inches (front/rear)
WHEELS	20-inches
TIRES	245/45ZR-20 (front), 275/40ZR-20 (rear)
CONSTRUCTION	unit body
SUSPENSION	multi-link strut, coil springs front/multi-link, coil springs rear
STEERING	rack and pinion
BRAKES	14.0-inch disc front/14.4-inch disc rear
ENGINE	312-horsepower, 3.6-liter V-6; 400-, 426-horsepower, 6.2-liter V-8
BORE AND STROKE	3.70 x 3.37 inches (3.6-liter), 4.06 x 3.62 inches (6.2-liter)
COMPRESSION	11.3:1 (3.6-liter), 10.7:1 (6.2-liter)
FUEL DELIVERY	SIDI (3.6-liter), SFI (6.2-liter)
TRANSMISSION	six-speed manual, six-speed automatic
AXLE RATIO	3.27:1 (3.6-liter), 3.45:1 (6.2-liter)
PRODUCTION	81,299

Given the long association between the Indianapolis 500 and Camaro pace cars, choosing the fifth-generation version as a pace car was a no-brainer. *General Motors 2012*

Some of the dealer-installed accessories in the series' Package A was a grille with painted body-color surround, painted ground effects, a body-color blade spoiler, and a red engine cover. Handsome 21-inch alloy wheels with Blade Silver Inserts and a Red Flange Stripe were part of the dealer installed Package B. Overall, the Synergy Series promised to deliver tasteful visual enhancements to an already handsome vehicle.

At the 2010 SEMA show in Las Vegas, Chevrolet unveiled a staggering track-only concept, called the Camaro SSX. Developed in conjunction with Riley Technologies and

Pratt & Miller, the SSX was designed to showcase the wide range of performance-enhancing parts available for the current-generation Camaro. With a full roll cage, carbon fiber hood, fenders, doors, adjustable rear spoiler, decklid, and a carbon fiber splitter cleaving the air, the SSX's 6.2-liter V-8 has been massaged with a special camshaft, cylinder heads and dry-sump oiling system to rip around a racetrack at crazy-fast speeds. Sounds like a lot of fun!

Surprises for 2012 include a return of the Z28, equipped with the 6.2-liter V-8, but wearing a supercharger. The Z28 was originally due in showrooms in 2010, but General Motors' bankruptcy put a wrench in the release date. This is the same engine used in Cadillac's successful CTS-V Series; and in that application, the brawny LSA powerplant generates an asphalt-wrinkling 556 horsepower. In the Corvette ZR1, this engine, denoted LS9, creates a staggering 620 horsepower As the Z28 will be at least 300 pounds lighter than the Cadillac, the performance potential is staggering. As the LSA engine can be used with both a manual and automatic transmission, and the LS9 can only be bolted to a manual, it's a fair assumption that the LSA powerplant will be slipped beneath the Z28's hood. Besides, the last thing Chevrolet wants to do is upstage the Corvette. While the Z28 will make mincemeat of a base Corvette, the top-dog ZR1 'Vette will be safe from the Z28. Huge disc brakes at each corner, quad tailpipes, and 20-inch wheels surrounded by huge, very sticky tires will be distinguishing cues that separate it from its more mainstream Camaro brethren. Targets for the Z28 include the Ford Shelby GT500 and the Dodge Challenger SRT-8, and with those vehicles on the market already, Chevrolet has had time to hone and prepare the Z28 for class-leading performance. Chassis tuning was carried out at the famed Nürburgring in Germany, where competence at speed is mandatory.

The convertible Camaro hit showrooms in the spring of 2011, just in time for the summer season. Farther downstream, the next generation of Camaro is already being created. It will roll on the Alpha platform, a structure more compact than the current Zeta base. Chevrolet has a staggering amount of Camaro heritage that it can draw upon, and a performance-hungry public is eating it up. It looks like the Camaro's best days are still to come!

Given the long history of Camaro Indianapolis pace-car replicas, it would have been shocking not to see this car sitting on this track. *General Motors 2012*

The convertible version of the new Camaro appeared for the 2011 model year. *General Motors 2012*

Opposite top: The Camaro convertible looked as badass with its top up as it did with the top down. *General Motors 2012*

Opposite bottom: The new Camaro looked even better topless. *General Motors 2012*

2011

In 2010, the Camaro outsold the Mustang 81,299 to 73,716 units. For 2011, Chevrolet introduced the convertible version of the Camaro, and the situation improved even more for Chevrolet. Camaro sales increased while Mustang sales decreased. Chevrolet sold 88,249 Camaros, and Ford sold only 70,438 Mustangs.

Performance-wise, it's hard to explain the difference. Ford introduced new drivetrains to compete with the Chevrolet, and by the numbers, the difference was minimal. For 2012, Ford introduced a new engine, the Coyote V-8, which displaced an iconic 302 cubic inches and cranked out 412 horsepower. While less than the LS3 engine in the Camaro SS, it had 200 fewer pounds to propel, giving it a horsepower-to-weight ratio that was roughly equal that of the Camaro. The performance of the two cars was so close that the abilities of the drivers accounted for more than the performance of the cars at stoplight drag races.

In base form, the Mustang's V-6 cranked out ten more horsepower than did the Camaro's overachieving V-6. When factoring in the weight difference, the base Mustang was actually faster than the base Camaro.

So why did Chevrolet win the sales war?

Most likely the difference came down to style. Not to take anything away from the Mustang, but the Camaro was simply the fresher of the two designs. Like the Mustang, it harkened back to the classic Camaros of the 1960s, but thanks to Ed Welburn's direction, it did so with a decidedly modern flare. The modern Camaro was unmistakably the descendant of the original car, but it was also a thoroughly modern design, one that could hold its own with anything else produced in the twenty-first century. By contrast, the 2010 redesign of the Mustang was really just the same old, same old, and the market reacted accordingly.

The introduction of the convertible also helped. While convertible sales accounted for only a small percentage of the cars when new, when the Camaro became a collectible car, the convertible version was the one everyone coveted. Not only did this drive up the price of used convertibles, but it also drove the sales of new convertibles.

Above: An icon reborn for the twenty-first century.
General Motors 2012

2011

MODEL AVAILABILITY	two-door coupe, two-door convertible
WHEELBASE	112.3 inches
LENGTH	190.4 inches
WIDTH	75.5 inches
HEIGHT	54.2 inches
WEIGHT	3,750 lbs
PRICE	$31,040 (SS)
TRACK	63.7/63.7 inches (front/rear)
WHEELS	20-inches
TIRES	245/45ZR-20 (front), 275/40ZR-20 (rear)
CONSTRUCTION	unit body
SUSPENSION	multi-link strut, coil springs front/multi-link, coil springs rear
STEERING	rack and pinion
BRAKES	14.0-inch disc front/14.4-inch disc rear
ENGINE	312-horsepower, 3.6-liter V-6; 400-, 426-horsepower, 6.2-liter V-8
BORE AND STROKE	3.70 x 3.37 inches (3.6-liter), 4.06 x 3.62 inches (6.2-liter)
COMPRESSION	11.3:1 (3.6-liter), 10.7:1 (6.2-liter)
FUEL DELIVERY	SIDI (3.6-liter), SFI (6.2-liter)
TRANSMISSION	six-speed manual, six-speed automatic
AXLE RATIO	3.27:1 (3.6-liter), 3.45:1 (6.2-liter)
PRODUCTION	88,249

Above: While the original Indianapolis Pace Car featured a big-block engine, the LS3 in the modern interpretation trumped the original by 60 horsepower. *General Motors 2012*

Left: The white/orange stripe exterior paint scheme was carried through to the interior in the 2011 pace car edition. Heavily bolstered seats held the driver and front passenger in place during spirited maneuvers. If you look closely, you can see the heads-up display unit atop the dashboard in front of the driver.

Below: With 426 horsepower, the 2011 Camaro Indy Pace Car didn't have any problems getting up to, and staying at, the speeds necessary to lead an angry pack of open-wheel race cars around the Brickyard. Unlike in years past, the powerplant in the 2011 Camaro was bone-stock—no need to massage it to generate enough grunt for the job.

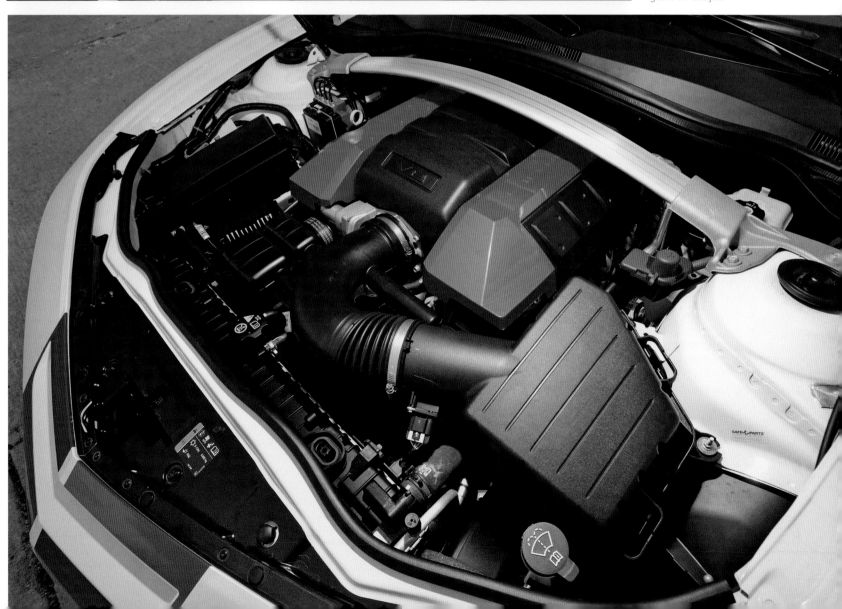

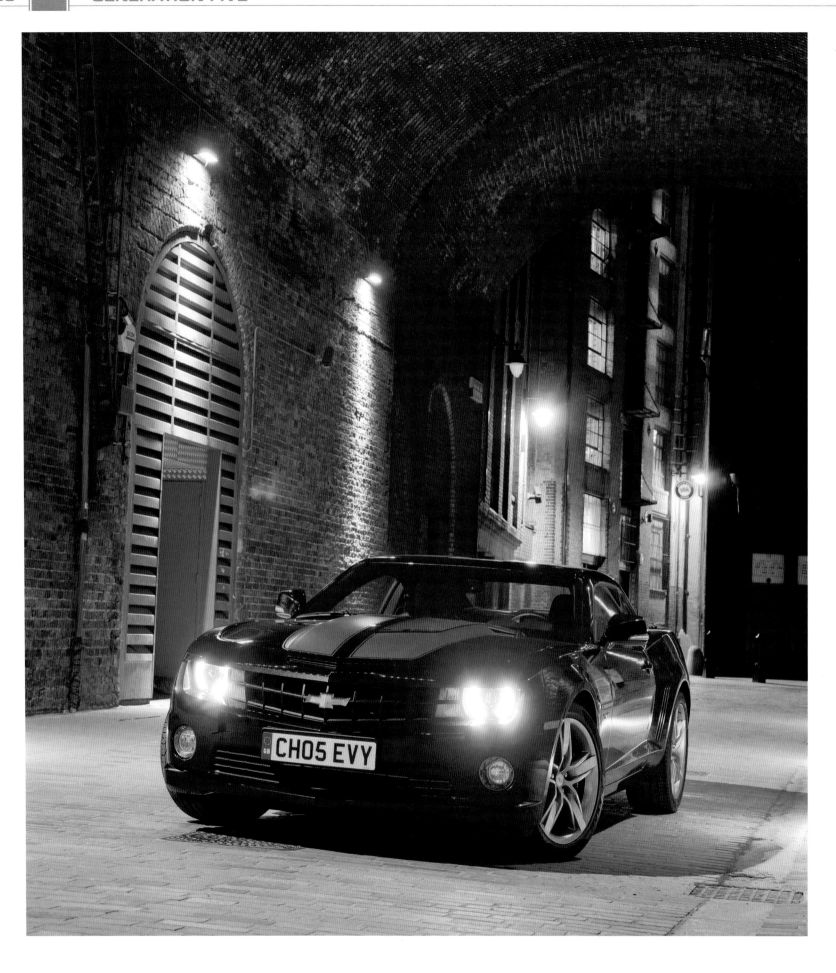

2012

In 2012, the sales battle tilted even more toward the Camaro. Chevrolet did not rest on its laurels for model year 2012. It pumped up the power of the base V-6 to 323 horsepower, handily trumping the Mustang's base engine.

The SS model didn't receive a power boost, but it did receive a new FE4 suspension and a new steering wheel to replace the much-maligned original. The big news, however, was the introduction of the ZL1. This was nothing short of the highest-performance Camaro ever built. Previously, that honor had been held by the limited-production race special built in 1969, which also bore the moniker ZL-1. This was COPO (Central Office Production Order—a program originally designed to provide special

order vehicles for fleet use, such as taxicabs or police vehicles, but subverted to provide vehicles for racing use to circumvent General Motors' racing ban in the 1960s) 9560, a full-blown drag-racing car with an all-aluminum 427-cubic-inch big-block engine. Chevrolet produced 69 of these monsters in 1969, and each fetches well over $1 million in today's market.

The new ZL1 will handily spank any of these million-dollar muscle cars on the race track. The original ZL-1 was rated at 425 horsepower, though actual output was probably closer to 500 horsepower. The supercharged V-8 engine in the 2012 ZL1 Camaro puts out 550 horsepower and 550 pound-feet of torque, making it the most potent engine ever mounted in a Camaro.

Above: Fans were grateful that the Camaro was back. The gorgeous 45th anniversary model was icing on the cake. *General Motors 2012*

Opposite: Not only did the new Camaro dominate at the track, it trumped the Mustang in the sales race too. *General Motors 2012*

2012

MODEL AVAILABILITY	two-door coupe
WHEELBASE	112.3 inches
LENGTH	190.4 inches
WIDTH	75.5 inches
HEIGHT	54.2 inches
WEIGHT	3750 lbs
PRICE	$31,040 (SS)
TRACK	63.7/63.7 inches (front/rear)
WHEELS	20-inches
TIRES	245/45ZR-20 (front), 275/40ZR-20 (rear)
CONSTRUCTION	unit body
SUSPENSION	multi-link strut, coil springs front/multi-link, coil springs rear
STEERING	rack and pinion
BRAKES	14.0-inch disc front/14.4-inch disc rear
ENGINE	323-horsepower, 3.6-liter V-6; 550-horsepower, 6.2-liter supercharged V-8
BORE AND STROKE	3.70 x 3.37 inches (3.6-liter), 4.06 x 3.62 inches (6.2-liter)
COMPRESSION	11.3:1 (3.6-liter), 10.7:1 (6.2-liter)
FUEL DELIVERY	SIDI (3.6-liter), SFI (6.2-liter)
TRANSMISSION	six-speed manual, six-speed automatic
AXLE RATIO	3.27:1 (3.6-liter), 3.45:1 (6.2-liter)
PRODUCTION	na

Above and below: Chevrolet celebrated the 45th anniversary of the Camaro in style. *General Motors 2012*

The new ZL1 uses an intercooled Eaton supercharger to generate this astonishing level of power. To cope with this prodigious output, every part of the car was beefed up. A special version of the standard Tremec TR-6060 six-speed manual transmission transmits the power in conjunction with a beefed-up driveshaft and a dual-mass flywheel to get all those horses to the rear wheels. Beefed-up suspension and brake components help keep the insane powertrain under control.

2013

With the most powerful Camaro ever built in its lineup, one might have expected Chevrolet to take a break. One would have been wrong. For 2013, Chevrolet reintroduced another iconic Camaro, the 1LE. This refers to the racing package originally introduced in the 1980s. The modern version retains the LS3 V-8 (most sane people realize that 435 horsepower is more than enough on any track) but utilized an improved suspension. There was no question that the 1LE package's sole reason for being was to humiliate Ford's Boss 302 Mustang, and the 1LE had everything it needed to do just that.

With the release of the fifth-generation Camaro for the 2010 model year, it was clear that General Motors was serious about competing in the muscle car arena, again.

Above: If the 435 horsepower of the LS3 SS wasn't enough, in 2012, buyers could select the insanely powerful ZL-1 version. *General Motors 2012*

Left: With 550 horsepower and 550 lb-ft of torque, the ZL1 Camaro shames even the original COPO ZL-1 of 1969. *General Motors 2012*

2013

MODEL AVAILABILITY	two-door coupe, convertible
WHEELBASE	112.3 inches
LENGTH	190.6 inches
WIDTH	75.5 inches
HEIGHT	54.2/54.7 coupe/convertible
WEIGHT	3,712 lbs
PRICE	$23,345
TRACK	63.7/64.1 inches (front/rear) LS & LT, 63.7/63.7 inches (front/rear) SS & ZL1
WHEELS	18 x 7.5 inches LS & 1LT, 19 x 8 inches 2LT, 20 x 8/20 x 9 inches (front/rear) SS, 20 x 10/20 x 11 inches (front/rear) SS w/1LE & ZL1
TIRES	P245/55R18 LS & 1LT, P245/50R19 2LT, P245/45R20/P275/40R20 (front/rear) LT w/RS package, P245/45R20/P275/40R20 SS, P285/35ZR20 1LE, P285/35ZRR20/P305/35ZR20 (front/rear) ZL1
CONSTRUCTION	unit body
SUSPENSION	double-ball-joint, multi-link strut/4.5-link independent (front/rear)
STEERING	rack and pinion, electric variable on SS
BRAKES	12.64/12.4 inches (front/rear) LS & LT, 14.0/14.4 inches (front/rear) SS & 1LE, 14.6/14.4 inches (front/rear) ZL1
ENGINE	323-horsepower 3.6-liter V-6, 426-horsepower 6.2-liter V-8, 580-horsepower 6.2-liter V-8 (ZL1)
BORE AND STROKE	3.70 x 3.37 inches (3.6-liter), 4.06 x 3.62 inches (6.2-liter)
COMPRESSION	11.3:1 (3.6-liter), 10.7:1 (6.2-liter with manual & 1LE), 10.4:1 (SS w/ auto), 9.1:1 (ZL1)
FUEL DELIVERY	direct fuel injection (3.6-liter), sequential port fuel injection (6.2-liter)
TRANSMISSION	six-speed manual, six-speed automatic
AXLE RATIO	3.27:1 (Hydra-Matic & Aisin), 3.45:1 (SS w/manual), 3.23:1 (ZL1 w/ automatic), 3.73:1 (ZL1 w/manual)
PRODUCTION	94.432

Chevrolet and toy maker Mattel teamed up to build 1,524 Hot Wheels Edition Camaros for model year 2013. Coupe production was 1,210 units, while convertibles totaled 314 built.

Brembo brakes lived inside the huge 21-inch alloy wheels fitted on to the 2013 Camaro Hot Wheels Edition. This is the same brake package found on regular Camaro SS models; the Hot Wheels Edition was strictly an appearance package, as no mechanical upgrades were used.

Chevrolet offered a Hot Wheels package for both coupe and convertible Camaros in 2013. The package cost $6,995 and included Kinetic Blue Metallic paint, a ZL1-sourced upper front grille and rear spoiler, and an RS Appearance Package. Huge 21-inch alloy wheels with a red stripe were part of the option.

Chevrolet targeted the 2013 Camaro 1LE to drivers that wanted to do battle with Ford's Mustang Boss 302. Unlike the Blue Oval's entry, which used a solid rear axle to keep the rear bumper off the ground, the Camaro 1LE used an independent rear suspension with a 3.91:1 gearset. The wheels were borrowed from the ZL1.

A bully in bully clothes, the ominous-looking 2013 Camaro 1LE is aggression squared. Its matte-black hood helps reduce reflections in the driver's eyes when driving into the sun, and the wheelwell-filling rolling stock clings to the road like a hungry leech.

The 2013 Camaro ZL1's engine compartment is eye candy for enthusiasts, and its 580 horsepower and 55 lb-ft of torque is enough to hurl the car toward a top speed in the region of 195 miles per hour. The LSA engine was the same powerplant Chevrolet used in the Corvette ZR1 and Cadillac CTS-V.

For 2013, the Camaro ZL1 was the weapon that Chevrolet took to the Nürburgring Nordschleife racetrack in Germany to put the world on notice that the Camaro was a true world-class sports car. The ZL1 cracked out a lap in 7:41.27 minutes.

Bowtie-loving drivers responded by snapping up the new car in droves. With Chrysler bringing back the Challenger, the American automotive landscape looked like a rewind to 1970. Ford's Mustang was soldiering on with a live rear axle, but its days under the original Pony Car were numbered.

As the fifth-generation Camaro entered its mid-life cycle in 2013, Chevrolet continued to offer the car in six- and eight-cylinder configurations. The base 1LS engine was a 323-horsepower V-6 powerplant, and it was popular with buyers who wanted a healthy dose of grunt with fuel economy that was wallet-friendly. Camaro chief engineer Al Oppenheiser and his crew had been working on the buyer's behalf. Want a bit more thrust? Then the SS model, with its 6.2-liter V-8, was the choice. Rated at 400 horses, it delivered classic V-8 thrust and grumble. Not enough power? Then order the impressive ZL1, a supercharged LSA 6.2-liter hand grenade. With 580 horsepower and 556 lb-ft of torque, an aggressively driven ZL1 could make a tire dealer very happy.

Left: Virtually all of the driver touch points in the interior of the 2013 ZL1 are suede, known for its grip and comfort. In a vehicle designed for serious track work, and able to nudge 195 miles per hour, the last thing a driver wants is a steering wheel that is slippery.

Below: Leaving rubber on the road is never a challenge in a 2013 Camaro ZL1. Yet for a vehicle that is so comfortable on a racetrack, it's remarkable that it can be used as a daily driver. The secret is to keep your foot out of it, then it's as docile as a V-6. Okay, a V-6 with a lot of attitude.

Right: The LS3 6.2-liter engine in the 2014 Camaro Pace Car edition uses an aluminum-block, alloy L92-type rectangular-port heads, a high-lift hydraulic camshaft, and 10.7:1 compression to generate 400 horsepower.

Opposite, top: Chevrolet went for a low-key look for the 2014 1LE, if any Camaro can be thought of as low-key. The matte-black hood with functional vents was actually covered by a vinyl wrap developed by 3M. It was recommended by the manufacturer that only hand washing or touchless car washes be used.

Opposite, bottom: For the eighth time in the Camaro's history, it was picked in 2014 to lead the pack at the Indianapolis 500. To honor that feat, Chevrolet offered a Super Sport 2SS Pace Car edition, complete with the 6.2-liter V-8. The interior of the 2014 pace car edition was standard Camaro 2SS, which is a pretty good place to start. Unlike the first-generation Camaros, the 2014 version offered seats that were leather-covered, power-adjustable, and heated, and actually held occupants in place.

2014

Model year 2014 saw a mid-cycle freshening, with subtle massaging of the sheet metal, including a revised front end and new taillights. Not so subtle was the introduction of another storied name in Camaro lore—the Z/28. Under the hood was the engine from the outgoing Corvette Z06, an LS7 7.0-liter mill that delivered 505 horsepower and 470 lb-ft of twist. For the first time in Camaro history, a dry-sump lubrication system was standard. Available only with a six-speed manual transmission, it enjoyed special suspension tuning, and huge carbon ceramic Brembo brakes. Nineteen-inch wheels were fitted with Pirelli Trofeo R tires. Like the original Z/28, the 2014 iteration was essentially a track car with a license-plate frame. Chevrolet didn't offer the Z/28 option with a convertible; for a performance car, the key is structural rigidity, and the coupe had that in spades. The ragtop used stiffeners to brace the body, and they added weight, the enemy of speed. In an effort to pare down the weight further, Chevrolet removed most of the sound-deadening material in the Z/28 and used thinner glass on the rear quarter windows. Evidently, speakers weigh a lot; the Z/28 used just one for the doorbell chime. Air conditioning, usually standard in Camaros, was an option on the Z/28.

If you wanted a Camaro with impressive track-like manners but didn't want to pay for the Z/28's lofty insurance premiums, then the hot ticket was to order the Camaro SS 1LE. This $3,500 option, which debuted in the 2013 model year, might have been the best money you could spend in a vehicle. The LS3 6.2-liter V-8 was unchanged, but the suspension was essentially lifted from the ZL1, giving the Camaro uncanny levels of handling performance. Underpinning changes included replacing the standard SS dual-tube shock absorbers with mono-tube units, beefier front and rear anti-roll bars, front shock tower braces, short-throw shifter, and wheels enlarged to 20-inch and widened to 10-inch in front, 11-inch in the rear. These huge aluminum wheels were surrounded by Goodyear Eagle F1 SuperCarG:2 tires sourced from the front of the ZL1. Oh, and the fuel pump with multiple pickup points, the wheel bearings, and the rear shock mounts were borrowed from the ZL1 as well. The rear axle ratio of 3.91:1 meant that off-the-line acceleration was more than a little noteworthy. Want all these greasy bits and an automatic transmission too? Sorry, Sparky! Tremec TR6060-MM6 manual transmission only.

The result was a grin-generating machine of the first order. Ride quality was a bit stiff, as the electric power

2014

MODEL AVAILABILITY	two-door coupe, convertible
WHEELBASE	112.3 inches
LENGTH	190.6 inches
WIDTH	75.5 inches
HEIGHT	54.2/54.7 coupe/convertible
WEIGHT	3,702 lbs
PRICE	$23,555
TRACK	63.7/64.1 inches (front/rear) LS, LT; 63.7/63.7 inches (front/rear) SS
WHEELS	18 x 7.5 inches LS & 1LT, 19 x 8 inches 2LT, 20 x 8/20 x 9 inches (front/rear) SS, 20 x 10/20 x 11 (front/rear) 1LE, ZL1
TIRES	P245/55R18 LS, 1LT; P245/50R19 1LT, 2LT; P245/45R20 RS; P245/45R20/P275/40R20 (front/rear) SS; P285/35ZR20 SS w/1LE; P285/35ZR20/P305/35ZR20 (front/rear) ZL1
CONSTRUCTION	unit body
SUSPENSION	double-ball-joint, multi-link strut/4.5-link independent (front/rear)
STEERING	rack and pinion, electric variable on SS
BRAKES	12.64/12.4 inches (front/rear) LS & LT; 14.0/14.4 inches (front/rear) SS; 14.6/14.4 (front/rear) ZL1
ENGINE	323-horsepower 3.6-liter V-6, 426-horsepower 6.2-liter V-8, 580-horsepower 6.2-liter V-8 (ZL1), 505-horsepower 7.0-liter V-8 (Z28)
BORE AND STROKE	3.70 x 3.37 inches (3.6-liter), 4.06 x 3.62 inches (6.2-liter), 4.125 x 4.00 inches (7.0-liter)
COMPRESSION	11.3:1 (3.6-liter), 10.7:1 (6.2-liter SS w/manual & 1LE), 10.4:1 (SS w/auto), 9.1:1 (ZL1), 11.0:1 (Z28)
FUEL DELIVERY	direct fuel injection (3.6-liter), sequential port fuel injection (6.2-liter, 7.0-liter)
TRANSMISSION	6-speed manual, 6-speed automatic
AXLE RATIO	3.27:1 (Hydra-Matic & Aisin), 3.45:1 (SS w/manual), 3.23:1 (ZL1 w/automatic), 3.91:1 (ZL1 w/manual)
PRODUCTION	97,632

Above: The huge chin air splitter on the 2014 Z/28 wasn't there for looks; on a racetrack, it helps direct air away from the underside of the vehicle for better aerodynamics. Chevrolet made no effort to portray the Z/28 as anything but a track car that you can drive on the street.

Right: Seven liters of mechanical mischief. The 2014 Z/28 was pretty much a track car with a license-plate holder. The port-injected LS7 engine, "borrowed" from the Corvette Z06, uses a dry-sump lubrication system to help its 505 horsepower live in cornering situations that might result in oil starvation in lesser cars. Only one transmission was available in the 2014 Z/28: a six-speed manual mated to a 3.91:1 rear end gear. This is the same transmission found in the Camaro SS 1LE.

MODEL AVAILABILITY	two-door coupe, convertible
WHEELBASE	112.3 inches
LENGTH	190.6 inches
WIDTH	75.5 inches
HEIGHT	54.2/54.7 coupe/convertible
WEIGHT	3702 lbs
PRICE	$23,705
TRACK	63.7/64.1 inches (front/rear) LS, LT; 63.7/63.7 inches (front/rear) SS
WHEELS	18 x 7.5 inches LS & 1LT, 19 x 8 inches 2LT, 20 x 8/20 x 9 inches (front/rear) SS, 20 x 10/20 x 11 inches (front/rear) 1LE
TIRES	P245/55R18 LS, P245/55R18, P245/50R19 LT, P245/45ZR20/P275/40ZR20 (front/rear) SS, P285/35ZR20 (front/rear) 1LE
CONSTRUCTION	unit body
SUSPENSION	double-ball-joint, multi-link strut/4.5-link independent (front/rear)
STEERING	rack and pinion, electric variable on SS
BRAKES	12.64/12.4 inches (front/rear) LS & LT, 14.0/14.4 inches (front/rear) SS
ENGINE	323-horsepower 3.6-liter V-6, 426-horsepower 6.2-liter V-8
BORE AND STROKE	3.70 x 3.37 inches (3.6-liter), 4.06 x 3.62 inches (6.2-liter)
COMPRESSION	11.3:1 (3.6-liter); 10.7:1 (LS3); 10.4:1 (L99)
FUEL DELIVERY	direct fuel injection (3.6-liter), sequential fuel injection (6.2-liter)
TRANSMISSION	6-speed manual, 6-speed automatic
AXLE RATIO	2.92:1 (2LS), 3.27:1 (Hydra-Matic & Aisin), 3.45:1 (SS), 3.91:1 (SS w/1LE)
PRODUCTION	77,502

steering and aggressive suspension worked to feed the driver with an endless stream of road information. But buyers didn't spring for the 1LE option for a plush ride. With its V-8 bellowing, the 1LE could reach 60 miles per hour in just 4.5 seconds. On a drag strip, in bone-stock condition, the 1LE would trip the lights in 12.9 seconds at 111 miles per hour. This was a Camaro that enabled owners to make new friends, specifically in law enforcement.

2015

Model year 2015 was basically a carry-over from 2014. With a milieu of Camaros ranging from a commute-friendly V-6 to a street-legal race car that would rip your head off, Chevrolet felt that there was plenty of excitement in the lineup. What many people forget is that technology essentially saved the muscle car. The 2015 Camaro equipped with a V-6 engine can get 30 miles per gallon on the highway, yet, as *Kelly Blue Book* points out, it has 53 more horsepower than the 1979 Z28. That was the performance flagship that year, and the six-cylinder Camaro built in 2015 can dust the Z28 without breaking a sweat.

With a price of $3,500, the 1LE option for the 2015 Camaro SS was some of the best money you could spend for sheer bang-for-the-buck fun. With 11-inch wide wheels in the rear, a lot of rubber contacts the road, which is handy with a 426-horsepower, 6.2-liter V-8 rotating said wheels.

A single paint hue was added in 2015: Blue Velvet Metallic. Minor upgrades to the navigation and audio systems were the extent of the interior changes. For years, complaints had been targeting the low-rent interior, with its hard plastics and difficult-to-read gauges. While nothing in that department had changed for 2015, it was hoped that the next model would address the concerns. Headquarters had already green-lighted the next-generation Camaro, and they felt that enough resources had been poured into the fifth-generation car. Now the goal was to reimagine the Camaro for the sixth time.

The interior of the 2015 Camaro Green Flash Special Edition was available in any color, as long as it was Jet Black interior trim, with a Graphite Silver instrument panel and door inserts with Ice Blue light piping.

Right: Transmission choices in the 2015 Camaro Green Flash were one automatic and one manual. Either one was bolted to a 6.2-liter V-8. With the automatic, slipping the selector into M allowed the driver to use the shifting paddles on the steering wheel to control gear selection.

Far right: Chevrolet called this wheel design a split-spoke, but whatever the name, the 21-inch alloy wheel on the 2015 Camaro Green Flash is a superb design. They were 8.5 inches wide in the front, 9.5 inches wide in the rear.

Left: Cyber Gray racing stripes on the Emerald Green Metallic paint are a classy combination, and the 2015 Green Flash special-edition Camaro wore it well. Only 811 Camaros were ordered with the option, 729 coupes and 82 convertibles. With its Emerald Green Metallic paint and chrome dusk 21-inch alloy wheels, the 2015 Camaro Green Flash Edition was a striking package. It was only fitted on 2SS models; both automatic and manual transmissions were available.

Below: The 21-inch wheels on the 2015 Camaro Green Flash special-edition did a nice job filling the wheel arches. LED taillights were part of the RS Appearance Package, as was the shark-fin roof antenna.

6

GENERATION SIX

2016–Today

Opposite: "Lean Muscularity" was the form vocabulary that guided the exterior design of the 2016 Camaro SS. Taut lines flow from one surface to the next, making the sheet metal appear to be pulled tightly against the mechanical sub-structure. Headlight technology allowed for smaller lamp housings, yet the quality of road illumination is superior to preceding generations.

Geeneral Motors knew they had a good thing with the fifth-generation Camaro. It was outselling its chief rival, the Ford Mustang, by 10 percent each year, and the demand wasn't softening. But in the performance car segment, change happens quickly. Chevrolet knew that a new Mustang was in the pipeline, equipped with independent rear suspension, the available 5.0-liter V-8, and an unbroken lineage dating back to April 15, 1964. The next-generation Camaro didn't just have to be good—it had to make people get up and dance.

When automotive journalists from around the world filed into Detroit's Belle Isle Park on Saturday, May 16, 2015, they were presented with a staggering display of Camaros, from the first one built to the latest test mules. Various department heads in the Camaro team spoke about the newest iteration before the car was unveiled. The new Camaro was not just revolutionary—it was *evolutionary*, and it would change the way performance models were viewed. It was clear that the key visual cues that screamed Camaro were alive and well, including the long hood/short deck proportions, the handsome roof line, and an

When the original Camaro debuted in 1966, Chevrolet managers said that the word *Camaro* meant a small, vicious animal that ate Mustangs. With the release of the sixth-generation version in 2016, the Camaro shed weight and continued to hunt them. Tauter and more chiseled than its predecessor, more than 70 percent of the vehicle content is unique to the car and not shared with anything else in the GM inventory.

aggressive front end. One of the complaints about the fifth-generation Camaros was their weight, and by using a new platform, that concern was history.

Chevrolet didn't waste any time popping champagne corks when the fifth-gen Camaro debuted; in the summer of 2009, the decision had already been made to create the sixth-generation Camaro. The competition in this segment was as intense as in the "Glory Days," and with the advent of computers in design and engineering, the time between initial corporate interest and Job One was considerably less than decades before. Al Oppenheiser was to be the chief engineer for the new car, and when he got the assignment, he was informed that the newest Camaro was going to sit on a new platform, riding atop the Alpha from Cadillac.

The fourth- and fifth-gen Camaros had been built in Canada. The sixth generation would return to the United States. General Motors announced on November 19, 2012, that the new Camaro would be built at the Lansing Grand River Assembly Plant, in Lansing, Michigan. This facility began operations in 2001 and builds vehicles on GM's Sigma and Alpha platforms.

A little background on the Alpha platform is in order. It was designed by engineers working under David Leone, and it is constructed of aluminum and high-strength steel with an eye to reducing weight. Designed for a multitude of uses, including front-wheel, all-wheel, and rear-drive applications, the Alpha platform is used in small and mid-sized vehicles. The first production car to use the Alpha platform was the 2013 Cadillac ATS. But the Camaro didn't have the proportions that the Cadillacs had—lower, wider, and with a long hood and short rear deck, the new Camaro couldn't come across as a two-door Cadillac.

Oppenheiser and his team set about transforming the standard Alpha platform into a proper starting point for the newest Camaro. Almost a year of work later, they had created just what the Camaro needed to retain the unique flavor of the marque while reducing weight at the same time. To reach the width mark that Tom Peters, leader of the Camaro design team, wanted, the suspension components were lengthened. Unlike the Cadillacs that used the Alpha platform, the fitment of a healthy V-8 in the Camaro had always been a given.

Left: Thick C-pillars have been a hallmark of Camaro exterior design since 1967, and they convey in a glance the arrival of the Chevrolet pony car. Though they create a heck of a blind spot, drivers quickly learn to work around them.

Below: Built on GM's Alpha platform, the 2016 Camaro SS is lithe and beautifully proportioned. Though the newest Camaro lost weight, approximately 200 pounds, the 6.2-liter LT1 V-8 found under the SS's hood lost none of its 455 horsepower. It makes for an agreeable power-to-weight ratio. When preliminary sketches for what would become the 2016 Camaro were needed, GM Design Chief Ed Welburn and studio head Tom Peters chose designer Hwasup Lee's proposal. Lee was working in GM's Warren Performance studio at the time and had been the lead exterior designer for the sixth-generation Corvette.

2016

MODEL AVAILABILITY	two-door coupe, convertible
WHEELBASE	110.7 inches
LENGTH	188.3 inches
WIDTH	74.7 inches
HEIGHT	53.1 inches
WEIGHT	3,685 lbs
PRICE	$26,695 (SS)
TRACK	63.0/62.9 inches (front/rear)
WHEELS	20x8.5/20x9.5 inches (front/rear)
TIRES	P245/40ZR20/P275/35ZR20 (front/rear)
CONSTRUCTION	unit body
SUSPENSION	strut/multilink (front/rear)STEERING rack and pinion
STEERING	rack and pinion, electric power
BRAKES	13.6-inch vented disc/13.3-inch vented disc (front/rear)
ENGINE	275-horsepower 2.0-liter I-4 turbo, 335-horsepower 3.6-liter V-6, 455-horsepower 6.2-liter V-8
BORE AND STROKE	3.39 x 3.39 inches (2.0-liter), 3.74 x 3.37 inches (3.6-liter), 4.06 x 3.62 inches (6.2-liter)
COMPRESSION	9.5:1 (2.0-liter), 11.5:1 (3.6-liter & 6.2-liter)
FUEL DELIVERY	direct high-pressure fuel injection
TRANSMISSION	6-speed manual, 8-speed automatic
AXLE RATIO	3.27:1 (2.0-liter & 3.6-liter manual), 3.27:1 (2.0-liter automatic), 2.77:1 (3.6-liter automatic), 3.73:1 (6.2-liter manual), 2.77:1 (6.2-liter automatic)

The 2016 Camaro SS uses horizontal daytime running lights (DRL) in its aggressive front end. Camaro RS models use tall, vertical DRLs. Hood vents are unique to the SS model.

2016

The newest Camaro had been put on a weight reduction program, and it showed. The fifth-gen Camaro, forever destined to compete with its arch-nemesis, the Mustang, had a problem overcoming its heft compared to its equine counterpart. Colin Chapman, founder of Lotus, coined the phrase, "adding lightness," meaning that reducing the weight of a car would benefit every performance parameter. By using the Alpha platform, Chevrolet "added lightness." And while the scales showed a 300-pound reduction, the engines in the newest Camaro were not delivering lower numbers—it's all about power-to-weight ratios, after all. The resulting package meant that the midnight oil was going to be burning in Dearborn.

The sixth-gen Camaro was going to utilize a four-cylinder engine for the first time in a long time. The last time a four-banger had been in a Camaro, the engine had been the Iron Duke, a cast-iron lump that laughed at any hint of performance or durability. Chevrolet had a trio of engines planned for the new Camaro: an I-4, a V-6, and a V-8. None of the engines in the latest Camaro were carried over from the fifth-generation. With the narrower engine compartment due to the Alpha platform, engineers had to get creative in

slipping the various powerplants between the front wheels. Each engine has its own strengths, however, that enhance the Camaro experience. Let's start our exploration of the new engines with the entry-level four-cylinder, which is a far cry from the small engines in Camaros of yore.

For most of its life, the Camaro has been sold in North America alone. General Motors wanted to change that and make the car available on the world stage. In many countries, a V-6, and especially a V-8, are taxed at a prohibitive level, making ownership of a four-cylinder powerplant almost mandatory. In the North American market, many younger drivers had been weaned on small-displacement four-cylinder engines, oftentimes using forced induction. So the idea of a (relatively) small engine in a Camaro RS was attractive to this desirable target audience. Chevrolet responded with the LTG 2.0-liter Ecotec I-4, rated at an attractive 275-horsepower and 295 lb-ft of torque. On a drag strip, this equates to 0 to 60 miles per hour in 5.2 seconds and a quarter-mile time of 13.9 seconds at 101 miles per hour. How does Chevy extract this kind of grunt from an inline four? The aforementioned forced induction is the key.

This engine is a member of a family of Gen III four-cylinder powerplants that General Motors debuted in 2013 and calls LE (Large Ecotec). It starts with the aluminum block and Rotacast head, displacing 121.9 cubic inches, or 1,998cc. It is fitted with dual overhead camshafts, balance shafts, forged connecting rods, sodium-filled exhaust valves, and direct-injection fuel delivery. Pushing more air into the combustion chambers is a twin-scroll turbocharger that uses an electronically controlled wastegate/bypass valve. The engine has a 9.5:1 compression ratio and requires premium unleaded fuel. Peak power is developed at 5,600 revolutions per minute, though the redline is 7,000 revs.

Camaro RSs for 2016 that use the Ecotec engine are labeled 1LT, while the next engine in the lineup is the V-6, and that is denoted as a 2LT Camaro. General Motors invested in an all-new engine for the V-6 powered Camaro, and it's a doozy. Displacing 3,649cc, the LGX V-6 is an all-aluminum powerplant, using dual overhead camshafts, and rated at 335 horsepower at a lofty 6,800 rpm and maximum torque of 284 lb-ft at 5,300 rpm. The engine's redline is 7,200 rpm and it uses a 11.5:1 compression ratio, yet regular unleaded fuel is recommended.

Left: The interior design manager for the 2016 Camaro was Ryan Vaughn, who wore the same hat with the current generation Corvette. Inspired by the 1967 Camaro's interior, especially in front of the driver, the newest Camaro is very user friendly, yet with a modern flair. Note the absence of the small gauges in front of the shifter; climate controls now occupy that real estate.

Below: While the Gen-5 LT1 V-8 under the hood of the 2016 Camaro SS displaces 6.2 liters, it's considered a small-block. However, the aluminum powerplant doesn't generate small numbers—455 horsepower and 455 lb-ft of torque.

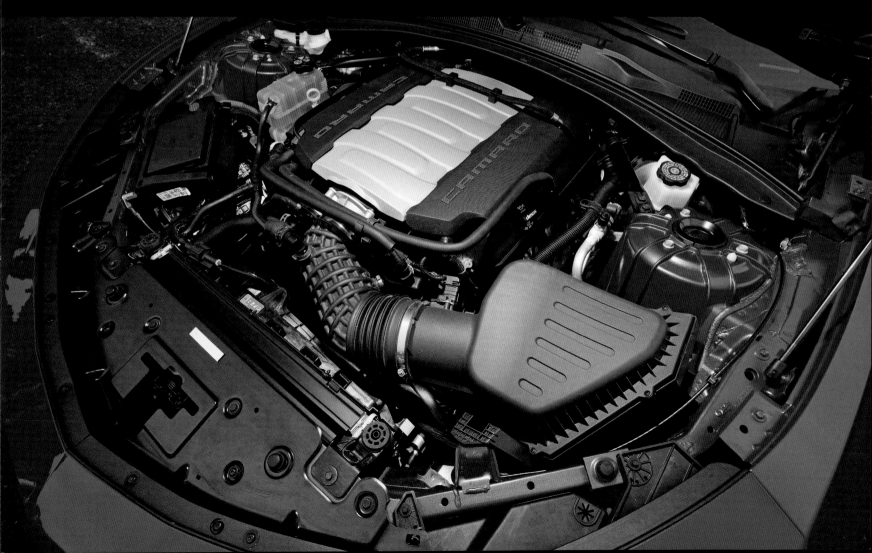

In Larry Edsall's excellent book *Camaro 2016*, design honcho Tom Peters talked about the newest Camaro's front end: "The narrow eyes and the large mouth add to the aggressive street fighter nature. By opening the lower grille, this provided more airflow, better engine cooling, which provides improved performance."

No longer is there enough room between the top of the tire and the wheel arch to stuff a medium-sized pet; the 20-inch alloy wheels perfectly fill the openings. The full-width rear spoiler is standard on the SS version.

Six-cylinder engines have come a long way since the debut of the 1967 Camaro, and the V-6 powerplant used in the 2016 Camaro is an all-new 3.6-liter design, using direct fuel injection, variable valve timing, and dual overhead camshafts to deliver 335 horsepower on regular unleaded fuel. This engine will vault the Camaro to 60 miles per hour in just 5.1 seconds.

In a first for General Motors on a dual overhead camshaft powerplant, this engine uses Active Fuel Management to cut fuel flow to two cylinders under light throttle, improving fuel mileage by up to 9 percent. Chevrolet claims that with this engine, the sixth-gen Camaro will vault to 60 miles per hour in just 5.1 seconds. While this engine is used in the 2016 Cadillac CTS, CT6, and ATS, in those vehicles a start/stop feature is used. Not so on the Camaro. Drivers of performance cars like to play with the throttle at stoplights, so Chevrolet lets them.

Speaking of performance, the Camaro's top-range engine, the LT1 6.2-liter V-8, oozes performance. The Gen V small-block saw duty in the 2014 Corvette, and Chevrolet didn't waste any time slipping it into the newest Camaro. It wasn't as easy as just opening the hood and dropping the LT1 into the space between the front wheels, though—with the narrow Alpha architecture, quite a few external changes had to be made to the engine to coax it into place. The most evident is the use of new exhaust manifolds. These aren't like

Above: Chevrolet installed modular underbody bracing to give the sixth-generation Camaro convertible impressive structural stiffness. From the beginning of the design process, the Camaro team planned to create a convertible version that could stand next to the coupe for lack of cowl shake and a minimization of noise, vibration, and harshness (NVH).

Right: The sixth-generation Camaro is available in coupe or convertible configurations, with the full range of available powerplants in both versions. This 2016 Camaro LT with the RS Appearance Package is fitted with the V-6 engine, vibrant Hyper Blue Metallic finish, and available 20-inch low-gloss black wheels. It is ready for a sojourn in the sun.

To access the wide-open spaces in a 2016 Camaro convertible, a single button is pushed. No more releasing latches above the windshield; the power-operated top is fully automatic. In fact, it can be cycled remotely using the key fob.

cast-iron manifolds of old, but rather, these are hand-built tubular headers.

Between the headers is an advanced V-8 that uses a plethora of racing technology to produce reliable power all day long. The engine uses Active Fuel Management to cut fuel to cylinders under low loads. Using the same 90-degree cylinder spacing and 4.400-inch bore center spacing as the first Chevrolet small-block V-8 from the 1950s, the Gen V engine is now built using extensive amounts of aluminum. A variable displacement oil pump ensures there's a constant stream of lubricant regardless of use. Unlike the dry sump system this engine uses in the Corvette, the direct-injection LT1 in the Camaro has a wet sump system. With 10 quarts, multiple oil pickup points, and a performance windage tray, oil starvation at high-g loads is a non-issue. The new Camaro SS was designed to generate 1.0 lateral g forces, and an engine that loses oil pressure under that stress quickly becomes a lump of inert metal.

Regardless of engine, getting the power to the ground is key to a great driving experience. Let's run through the transmissions that are used with each engine. Unlike many vehicles that have eschewed manual transmission for the

"convenience" of an automatic, the new Camaro offers both. New with the sixth-generation Camaro is an eight-gear automatic transmission. The new automatic is lighter than the older six-speed, thanks to extensive use of lightweight materials such as aluminum and magnesium. The SS is equipped with a GM 8L90 automatic, while the V-6 and turbo-4 employ the GM 8L45.

Manual transmission fans had plenty to love in the newest Camaro, as every engine available could be fitted with three pedals and a stick. In the SS, the Tremec TR6060 is used. This same box is used in the Corvette, filled with seven gears. In Camaro guise, it has six gears. But just like in the Corvette, the Tremec unit uses Active Rev Matching to match engine and transmission revolutions to provide silken up and downshifts. A twin-disc clutch (240mm) works in conjunction with the new manual transmission. The result is a refined feel when shifting as well as a diminishment of noise and vibration.

Drivers of either turbo-4 or V-6 Camaros equipped with a manual transmission stir a Tremec TR3160. This advanced transmission is a durable component and has an impressive list of features that increase longevity and deliver smooth operation. Double- and triple-cone synchronizers, high-strength steel on all gears and shafts, and high-capacity tapered bearings contribute to a transmission that can handle the day-to-day operation of a commuter vehicle.

2017

In the realm of commuting Camaros, Chevrolet brought back the ZL1 for 2017, and it would make short work of virtually any commute. If 650 horsepower and 650 lb-ft of torque don't help in merging onto the freeway, it's time to take the bus. As it has done in the past with its performance vehicles, Chevrolet ran the new ZL1 around the Nürburgring Nordschleife, and when the dust had settled, it had circled the "Green Hell" 12 seconds quicker than the previous generation ZL1. The YouTube video of this lap is impressive, as the car is essentially a street vehicle that hits 181 miles per hour on a straight, yet when the lap is over, the driver calmly uses the backup camera to park the car.

The ZL1 is quick—how does 0 to 60 in 3.5 seconds strike you? How about covering the quarter-mile in 11.4 seconds? Look out, curve ahead! Not to worry—the ZL1 can pull 1.02 gs all day long. Want to let the sunshine in? The 2017 Camaro ZL1 is available as a convertible as well as a coupe. Under the carbon fiber hood insert is the famed LT4 6.2-liter V-8, complete with an Eaton supercharger to accelerate the combustion process. Bolted onto the back of the engine are two choices of transmission, a six-speed manual and a ten-speed automatic.

More than 100 hours were spent in a wind tunnel honing the aerodynamic bits that help keep the ZL1 on the ground and the mechanical parts within recommended operating temperatures. The beast rides on massive 20-inch forged wheels, shod with Goodyear F1 SuperCar tires. Brembo brakes live at each corner, with 15.35-inch rotors in the front.

All of this race-honed technology doesn't come cheap; the base price on a 2017 ZL1 is $61,140. But when you realize what you get for the money, it's money well spent. Just keep the shiny side up.

The Camaro celebrates its 50th anniversary in 2017, and Chevrolet has never been shy about releasing anniversary models. Like on prior celebratory editions, special paint and graphics set it aside from the normal Camaros. Painted Nightfall Gray Metallic with orange accents, the 50th Anniversary Edition has plenty of unique badging, to celebrate fifty years of excellence and to honor the 1967 Chevrolet Camaro Indianapolis 500 Pace Car. The option is available on both 2LT and 2SS vehicles, coupe and convertible. Equipping the 2LT Camaro with the Anniversary package ($2,595)

The engine in this 2017 Camaro 2LT convertible is an Ecotec 2.0-liter inline-four-cylinder aluminum engine, rated at 275 horsepower and a hefty 295 lb-ft of torque—more torque than the V-6 engine! The Ecotec engine will work to get the Camaro to 60 miles per hour in just 5.4 seconds. EPA mileage is 31/22, highway/city.

requires the Convenience and Lighting package ($2,800), so it's going to cost $5,395 extra to fit the Anniversary bits onto the premium V-6 powered Camaro. Buying a 2SS Camaro? The price of the Anniversary package is $1,795.

The 1LE Track Package option returns to the Camaro lineup in 2017 and is available on V-6 Camaros for the first time. Sticky tires, a more aggressive suspension, improved braking, and special cooling aids help the 1LE survive the rigors of track use. The V-6 Camaro starts with the FE3 suspension, which consists of rear cradle mounts, upgraded 1LE anti-sway bars, firmer rear shocks, and ball-jointed rear toe links. Twenty-inch aluminum wheels are wrapped in Goodyear F1 tires, and Brembo 4-piston brakes in the front help to shed speed. A mechanical limited-slip differential is fitted with 3.27:1 gears.

When the 1LE option is installed on the 2017 Camaro SS, things get even more intense. Under the car, the FE4 suspension uses magnetorheological ride dampeners to keep wheel movements in check. Front brakes use six-point Brembos clamping huge 370mm rotors. The differential is electronically controlled, and forward of it is a six-speed

manual transmission. A short-throw shifter helps to snick into the next gear, and Recaro seats are standard on the SS 1LE, optional on the V-6 equipped Camaro. Speed on these beasts is limited to 155 miles per hour with the V-6, 186 miles per hour in the 1LE SS. That's off the showroom floor, in a car that you can live with every day. It's rather remarkable.

The future of the Camaro is assured. There will always be a place in some people's hearts for a vehicle that is fun to drive, looks great (and by extension, makes *them* look great), growls like an apex predator, can be used every day, and won't break the bank. This formula is just as true today as that day in September 1966 when the Chevrolet Camaro debuted. Pundits have been predicting the end of performance cars for decades, and today the vehicles are more thrilling than ever.

While Chevrolet hasn't announced it, whispers are swirling that the seventh-generation Camaro is already under development. Will it be exciting to drive and pleasing to the eye, and push all the right go-fast buttons? The answer to those questions will always be a resounding *YES*—long live the Camaro!

Above: It didn't take long for the ZL1 version of the sixth-gen Camaro to storm onto the road, as the 2017 model unleashed its 650 horsepower and 650 lb.-ft of torque onto an unsuspecting world. With its carbon fiber hood insert and imposing front end, it's a wolf in wolf's clothing. *GM*

Following page: The 2017 Camaro ZL1 is just as comfortable on the track as it is on the street. Its rear wheels measure a huge 20 inches in diameter, 11 inches width. And the sticky tires still have a problem getting all of the 6.2-liter LT4's power down. Chevrolet's MSRP is $61,140, and that sum will vault you to 60 mph in just 3.5 seconds. Sounds like good value. *GM*

SELECT OPTIONS

1967 OPTIONS

RPO	Descripton	Qty	$ Retail
12337	Camaro Sport Coupe, 6-cylinder	53,523	2,466.00
12367	Camaro Convertible, 6-cylinder	5,285	2,704.00
12437	Camaro Sport Coupe, 8-cylinder	142,242	2,572.00
12467	Camaro Convertible, 8-cylinder	19,856	2,809.00
L22	Engine, 250ci, 155hp Turbo-Thrift 6-cyl	38,165	26.35
L30	Engine, 327ci, 275hp, Turbo-Fire V8	25,287	92.70
L35	Super Sport, with 396ci, 325hp engine	4,003	263.30
L48	Super Sport, with 350ci, 295hp engine	29,270	210.65
L78	Super Sport, with 396ci, 375hp engine	1,138	500.30
Z22	Rally Sport Package	64,842	105.35
Z28	Special Performance Package (coupe)	602	358.10

1968 OPTIONS

RPO	Description	Qty	$ Retail
12337	Camaro Sport Coupe, 6-cylinder	47,456	2,565.00
12367	Camaro Convertible, 6-cylinder	3,513	2,802.00
12437	Camaro Sport Coupe, 8-cylinder	167,251	2,670.00
12467	Camaro Convertible, 8-cylinder	16,927	2,908.00
L22	Engine, 250ci, 155hp Turbo-Thrift 6-cyl	28,647	26.35
L30	Engine, 327ci, 275hp Turbo-Fire V8	21,686	92.70
L34	Super Sport, with 396ci, 350hp engine	2,579	368.65
L35	Super Sport, with 396ci, 325hp engine	10,773	263.30
L48	Super Sport, with 350ci, 295hp engine	12,496	210.65
L78	Super Sport, with 396ci, 375hp engine	4,575	500.30
L89	Super Sport, 396ci, 375hp, aluminum heads	272	868.95
Z22	Rally Sport Package	40,977	105.35
Z28	Special Performance Package (coupe)	7,199	400.25

1969 OPTIONS

RPO	Description	Qty	$ Retail
12337	Camaro Sport Coupe, 6-cylinder	34,541	2,621.00
12367	Camaro Convertible, 6-cylinder	1,707	2,835.00
12437	Camaro Sport Coupe, 8-cylinder	190,971	2,727.00
12467	Camaro Convertible, 8-cylinder	15,866	2,940.00
JL8	Power Brakes, front and rear disc	206	500.30
LM1	Engine, 350ci, 225hp Turbo-Fire V8	10,406	52.70
L22	Engine, 250ci, 155hp Turbo-Thrift 6-cyl	18,660	26.35
L34	Engine, 396ci, 350hp Turbo-Jet V8 (SS)	2,018	184.35
L35	Engine, 396ci, 325hp Turbo-Jet V8 (SS)	6,752	63.20
L48	Engine, 350ci, 300hp Turbo-Fire V8 (base SS)	22,339	-
L65	Engine, 350ci, 250hp Turbo-Fire V8	26,898	21.10
L78	Engine, 396ci, 375hp Turbo-Jetd V8 (SS)	4,889	316.00
L89	Engine, 396ci, 375hp V8 w/alum heads (SS)	311	710.95
Z22	Rally Sport Package	37,773	131.65
Z27	Super Sport (SS) package	34,932	295.95
Z28	Special Performance Package (coupe)	20,302	458.15

1970 OPTIONS

RPO	Description	Qty	$ Retail
12387	Camaro Sport Coupe, 6-cylinder	12,578	2,749.00
12487	Camaro Sport Coupe, 8-cylinder	112,323	2,839.00
L34	Engine, 396ci, 350hp Turbo-Jet V8 (SS)	1,864	152.75
L65	Engine, 350ci, 250hp Turbo-Fire V8	34,780	31.60
L78	Engine, 396ci, 375hp Turbo-Jet V8 (SS)	600	385.50
Z22	Rally Sport Package	27,136	168.55
Z27	Super Sport Package	12,476	289.65
Z28	Special Performance Package	8,733	572.95

1971 OPTIONS

RPO	Description	Qty	$ Retail
12387	Camaro Sport Coupe, 6-cylinder	11,178	2,758.00
12487	Camaro Sport Coupe, 8-cylinder	13,452	2,848.00
LS3	Engine, 300hp, 396ci Turbo-Jet V8 (SS)	1,533	99.05
L65	Engine, 245hp 350ci Turbo-Fire V8	34,017	26.35
Z22	Rally Sport Package	18,404	179.05
Z27	Super Sport Package	8,377	313.90
Z28	Special Performance Package	4,862	786.75

1972 OPTIONS

RPO	Description	Qty	$ Retail
12387	Camaro Sport Coupe, 6-cylinder	4,821	2,729.70
12487	Camaro Sport Coupe, 8-cylinder	63,830	2,819.70
LS3	Engine, 396ci, 240hp Turbo-Jet V8 (SS)	970	96.00
L65	Engine, 350ci, 165hp Turbo-Fire V8	27,009	26.00
Z22	Rally Sport Package	11,364	118.00
Z27	Super Sport Package	6,562	306.35
Z28	Special Performance Package	2,575	769.15
Z87	Customer Interior	6,462	113.00

1973 OPTIONS

RPO	Description	Qty	$ Retail
1FQ87	Camaro Sport Coupe, 6-cylinder	3,614	2,732.70
1FQ87	Camaro Sport Coupe, 8-cylinder	60,810	2,822.70
1FS87	Camaro Type LT Coupe, 8-cylinder	32,327	3,211.70
L48	Engine, 350ci, 175hp Turbo-Fire V-8	13,220	102.00
L65	Engine, 350ci, 145hp Turbo-Fire V-8	50,262	26.00
Z22	Rally Sport Package	11,364	118.00
Z28	Special Performance Package	11,574	598.05

1974 OPTIONS

RPO	Description	Qty	$ Retail
1FQ87	Camaro Sport Coupe, 6-cylinder	22,210	2,827.70
1FQ87	Camaro Sport Coupe, 8-cylinder	79,835	3,039.70
1FS87	Camaro Type LT Coupe, 8-cylinder	48,963	3,380.70
LM1	Engine, 350ci, 160hp Turbo Fire V8	9,139	46.00
L48	Engine, 350ci, 185hp Turbo-Fire V8	20,520	76.00
Z22	Rally Sport Package	11,364	118.00
Z28	Special Performance Package	13,802	572.05

1975 OPTIONS

RPO	Description	Qty	$ Retail
1FQ87	Camaro Sport Coupe, 6-cylinder	29,749	3,553.05
1FQ87	Camaro Sport Coupe, 8-cylinder	76,178	3,698.05
1FS87	Camaro Type LT Coupe, 8-cylinder	39,843	4,070.05
G95	Rear Axle, highway	1,596	12.00
J50	Power Brakes	49,356	55.00
LM1	Engine, 155hp, 350ci Turbo-Fire V8	31,569	54.00
Z85	Rally Sport Package	7,000	238.00

1976 OPTIONS

RPO	Description	Qty	$ Retail
1FQ87	Camaro Sport Coupe, 6-cylinder	38,047	3,762.35
1FQ87	Camaro Sport Coupe, 8-cylinder	92,491	3,927.35
1FS87	Camaro Type LT Coupe, 8-cylinder	52,421	4,320.35
LM1	Engine, 350ci, 165hp Turbo-Five V8	56,710	85.00
Z85	Rally Sport Package	15,855	260.00

1977 OPTIONS

RPO	Description	Qty	$ Retail
1FQ87	Camaro Sport Coupe, 6-cylinder	131,717	4,113.45
1FS87	Camaro Type LT Coupe, 6-cylinder	72,787	4,478.45
1FQ87	Camaro Z28 Sport Coupe, 8-cylinder	14,349	5,170.06
LG3	Engine, 305ci, 145hp V8 (135hp in California)	147,173	120.00
LM1	Engine, 350ci, 170hp V8 (160hp in California)	40,291	210.00
Z85	Rally Sport Package	17,026	281.00

1978 OPTIONS

RPO	Description	Qty	$ Retail
1FQ87	Camaro Sport Coupe, 6-cylinder	134,491	4,414.25
1FQ87	Camaro Rally Sport Coupe, 6-cylinder	11,902	4,784.25
1FS87	Camaro Type LT Coupe, 6-cylinder	65,635	4,814.25
1FS87	Camaro Type LT Rally Sport Coupe, 6-cyl	5,696	5,065.25
1FQ87	Camaro Z28 Sport Coupe, 8-cylinder	54,907	5,603.85
LG3	Engine, 305ci, 145hp V8 (135hp in California)	143,110	185.00
LM1	Engine, 350ci, 170hp V8 (160hp in California)	92,539	300.00

1979 OPTIONS

RPO	Description	Qty	$ Retail
1FQ87	Camaro Sport Coupe, 6-cylinder	111,357	4,676.90
1FQ87	Camaro Rally Sport Coupe, 6-cylinder	19,101	5,072.90
1FS87	Camaro Berlinetta Coupe, 6-cylinder	67,236	5,395.90
1FQ87	Camaro Z28 Sport Coupe, 8-cylinder	84,877	6,115.35
LG3	Engine, 305ci, 145hp V8 (135hp in California)	138,197	235.00
LM1	Engine, 350ci, 170hp V8 (160hp in California)	122,461	360.00

1980 OPTIONS

RPO	Description	Qty	$ Retail
1FQ87	Camaro Sport Coupe, 6-cylinder	68,174	5,498.60
1FQ87	Camaro Rally Sport Coupe, 6-cylinder	12,015	5,915.60
1FS87	Camaro Berlinetta Coupe, 6-cylinder	26,679	6,261.60
1FQ87	Camaro Z28 Sport Coupe, 8-cylinder	45,137	7,121.32
LG4	Engine, 305ci, 155hp V8	47,580	295.00
LM1	Engine 350ci, 190hp V8 (included with Z28)	41,825	nc
L39	Engine, 267ci, 120hp V8	11,496	180.00

1981 OPTIONS

RPO	Description	Qty	$ Retail
1FP87	Camaro Sport Coupe, 6-cylinder	62,614	6,581.23
1FS87	Camaro Berlinetta Coupe, 6-cylinder	20,253	7,356.23
1FP87	Camaro Z28 Sport Coupe, 8-cylinder	43,272	8,025.23
LG4	Engine, 305ci, 155hp V8	17,909	75.00
LM1	Engine 350ci, 190hp V8 (included with Z28)	37,400	nc
L39	Engine, 267ci, 115hp V8	18,826	75.00

1982 OPTIONS

RPO	Description	Qty	$ Retail
1FP87	Camaro Sport Coupe, 4-cylinder	78,761	8,029.50
1FS87	Camaro Berlinetta Coupe, 6-cylinder	39,744	9,665.06
1FP87	Camaro Z28 Sport Coupe, 8-cylinder	64,882	10,099.26
1FP87	Camaro Z28 / Z50 Indy Edition, 8-cylinder	6.360	10,999.26
LC1	Engine, 173ci, 102hp V6 (std Berlinetta)	69,777	125.00
LG4	Engine, 305ci, 145hp V8 (std Z28)	73,495	295.00
LU5	Engine, 305ci, 165hp V8 (optional Z28 only)	24,673	450.00

1983 OPTIONS

RPO	Description	Qty	$ Retail
1FP87	Camaro Sport Coupe, 4-cylinder	63,806	8,036.00
1FS87	Camaro Berlinetta Coupe, 6-cylinder	27,825	9,881.00
1FP87	Camaro Z28 Sport Coupe, 8-cylinder	62,650	10,336.00
LC1	Engine, 173ci, 107hp V6 (std Berlinetta)	54,332	150.00
LG4	Engine, 305ci, 150hp V8 (std Z28)	67,053	350.00
LU5	Engine, 305ci, 175hp V8 (optional Z28 only)	19,847	450.00
L69	Engine, 305ci, 190hp V8 (optional Z28 only)	3,223	450.00

1984 OPTIONS

RPO	Description	Qty	$ Retail
1FP87	Camaro Sport Coupe, 4-cylinder	127,292	7,995.00
1FS87	Camaro Berlinetta Coupe, 6-cylinder	33,400	10,895.00
1FP87	Camaro Z28 Sport Coupe, 8-cylinder	100,899	10,620.00
1A3	Special Olympics Package	3,722	na
LC1	Engine, 173ci, 107hp V6 (std Berlinetta)	98,471	250.00
LG4	Engine, 305ci, 150hp V8 (std Z28)	99,976	550.00
L69	Engine, 305ci, 190hp V8 (optional Z28 only)	52,457	530.00

1985 OPTIONS

RPO	Description	Qty	$ Retail
x1FP87	Camaro Sport Coupe, 4-cylinder	97,966	8,363.00
1FS87	Camaro Berlinetta Coupe, 6-cylinder	13,649	11,060.00
1FP87	Camaro Z28 Sport Coupe, 8-cylinder	68,403	11,060.00
LB8	Engine, 173ci, 135hp V6 (std Berlinetta)	78,315	335.00
LB9	Engine, 305ci, 215hp V8 (Z28 only)	32,836	680.00
LG4	Engine, 305ci, 165hp V8 (std Z28)	63,052	635.00
L69	Engine, 305ci, 190hp V8 (IROC Z28 only)	2,497	680.00

1986 OPTIONS

RPO	Description	Qty	$ Retail
1FP87	Camaro Sport Coupe, 4-cylinder	99,608	9,349.00
1FS87	Camaro Berlinetta Coupe, 6-cylinder	4,479	12,316.00
1FP87	Camaro Z28 Sport Coupe, 8-cylinder	88,132	12,316.00
B4Z	IROC Sport Equipment Package	49,585	659.00
LB8	Engine, 173ci, 135hp V6 (std Berlinetta)	77,478	350.00
LB9	Engine, 305ci, 190hp V8 (Z28 only)	46,374	695.00
LG4	Engine, 305ci, 155hp V8 (std Z28)	68,293	750.00
L69	Engine, 305ci, 190hp V8 (Z28 only)	74	695.00

1987 OPTIONS

RPO	Description	Qty	$ Retail
1FP87	Camaro Sport Coupe, 6-cylinder	83,890	10,409.00
1FP87	Camaro Sport Coupe Convertible, 8-cylinder	263	15,208.00
1FP87	Camaro Z28 Sport Coupe, 8-cylinder	52,863	13,233.00
1FP87	Camaro Z28 Sport Coupe Convertible, 8-cyl	744	17,632.00
B2L	Engine, 350ci, 220hp V8 (IROC-Z only)	12,105	1,045.00
B4Z	IROC Sport Equipment Package	38,889	*
LB9	Engine, 305ci, 190hp (Z28 only)	28,370	745.00
LG4	Engine, 305ci, 165hp or 170hp (Sport coupe)	36,845	400.00

1988 OPTIONS

RPO	Description	Qty	$ Retail
1FP87	Camaro Sport Coupe, 6-cylinder	66,605	10,995.00
1FP87	Camaro Sport Coupe Convertible, 8-cylinder	1,859	16.255.00
1FP87	Camaro IROC-Z Coupe, 8-cylinder	24,050	13,490.00
1FP87	Camaro IROC-Z Convertible, 8-cylinder	3,761	18,105.00
B2L	Engine, 350ci, 220hp V8 (IROC-Z coupe only)	12,116	1,045.00
LB9	Engine, 305ci, 195hp (IROC-Z only)	12,620	745.00
L03	Engine, 305ci, 170hp (for Sport Coupe)	28,719	400.00

1989 OPTIONS

RPO	Description	Qty	$ Retail
1FP87	Camaro RS Coupe, 6-cylinder	83,487	11,495.00
1FP87	Camaro RS Convertible, 8-cylinder	3,245	16,995.00
1FP87	Camaro IROC-Z Coupe, 8-cylinder	20,067	14,145.00
1FP87	Camaro IROC-Z Convertible, 8-cylinder	3,940	18,945.00
1LE	Special Performance Components Package	111	*
B2L	Engine, 350ci, 220hp V8 (IROC-Z coupe only)	12,370	1,045.00
LB9	Engine, 305ci, 195hp (IROC-Z only)	8,925	745.00
L03	Engine, 305ci, 170hp (RS Coupe)	46,715	400.00

1990 OPTIONS

RPO	Description	Qty	$ Retail
1FP87	Camaro RS Coupe, 6-cylinder	28,750	10,995.00
1FP87	Camaro RS Convertible, 8-cylinder	729	16,880.00
1FP87	Camaro IROC-Z Coupe, 8-cylinder	4,213	14,555.00
1FP87	Camaro IROC-Z Convertible, 8-cylinder	1,294	20,195.00
1LE	Special Performance Components Package	62	*
B2L	Engine, 350ci, 245hp V8 (IROC-Z coupe only)	2,415	300.00
LB9	Engine, 305ci, 210hp (no cost with IROC-Z)	3,092	745.00
L03	Engine, 305ci, 170hp (for RS Coupe)	16,736	350.00

1991 OPTIONS

RPO	Description	Qty	$ Retail
1FP87	Camaro RS Coupe, 6-cylinder	79,854	12,180.00
1FP87	Camaro RS Convertible, 8-cylinder	5,329	17,960.00
1FP87	Camaro Z28 Coupe, 8-cylinder	12,452	15,445.00
1FP87	Camaro Z28 Convertible, 8-cylinder	3,203	20,815.00
1LE	Special Performance Components Package	478	*
B2L	Engine, 350ci, 245hp V8 (Z28 coupe only)	6,080	300.00
LB9	Engine, 305ci, 205hp (base with Z28)	9,996	nc
L03	Engine, 305ci, 170hp (for RS Coupe)	53,040	350.00

1992 OPTIONS

RPO	Description	Qty	$ Retail
1FP87	Camaro RS Coupe, 6-cylinder	60,994	12,075.00
1FP87	Camaro RS Convertible, 8-cylinder	2,562	18,055.00
1FP87	Camaro Z28 Coupe, 8-cylinder	5,197	16,055.00
1FP87	Camaro Z28 Convertible, 8-cylinder	1,254	21,500.00
1LE	Special Performance Components Package	705	*
B2L	Engine, 350ci, 245hp V8 (Z28 coupe only)	3,038	300.00
LB9	Engine, 305ci, 205hp (base with Z28)	4,002	nc
L03	Engine, 305ci, 170hp (for RS)	39,142	369.00

1993 OPTIONS

RPO	Description	Qty	$ Retail
1FP87	Camaro Coupe	21,253	13,399.00
1FP87	Camaro Z28 Coupe	17,850	16,779.00
1LE	Special Performance Components Package	19	310.00

1994 OPTIONS

RPO	Description	Qty	$ Retail
1FP87	Camaro Coupe	76,531	13,499.00
1FP67	Camaro Convertible	2,328	18,745.00
1FP87	Camaro Z28 Coupe	36,008	16,999.00
1FP67	Camaro Z28 Convertible	4,932	22,075.00
1LE	Special Performance Components Package	135	310.00

1995 OPTIONS

RPO	Description	Qty	$ Retail
1FP87	Camaro Coupe	77,431	14,250.00
1FP67	Camaro Convertible	6,948	19,495.00
1FP87	Camaro Z28 Coupe	30,335	17,915.00
1FP67	Camaro Z28 Convertible	8,024	23,095.00
1LE	Special Performance Components Package	106	310.00

1996 OPTIONS

RPO	Description	Qty	$ Retail
1FP87	Camaro Coupe	31,528	14,990.00
1FP67	Camaro Convertible	2,994	21,270.00
1FP87	Camaro Rally Sport Coupe	8,091	17,490.00
1FP67	Camaro Rally Sport Convertible	905	22,720.00
1FP87	Camaro Z28 Coupe	14,906	19,390.00
1FP67	Camaro Z28 Convertible	2,938	24,930.00
1LE	Special Performance Components Package	55	1,175.00
R7T	SS Package (non-Chevrolet modification)	2,257	3,999.00

1997 OPTIONS

RPO	Description	Qty	$ Retail
1FP87	Camaro Coupe	25,445	16.215.00
1FP67	Camaro Convertible	4,330	21,770.00
1FP87	Camaro Rally Sport Coupe	8,154	17,970.00
1FP67	Camaro Rally Sport Convertible	1,021	23,170.00
1FP87	Camaro Z28 Coupe	17,955	20,115.00
1FP67	Camaro Z28 Convertible	3,297	25,520.00
1LE	Special Performance Components Package	48	1,175.00
AU0	Remote Keyless Entry	44,921	225.00
AU3	Power Door Locks	49,632	220.00
A31	Power Windows *	46,493	275.00
B4C	Special Service Package ($3,454 w/automatic)	253	2,474.00
R7T	SS Package (non-Chevrolet modification)	3,137	3,999.00

1998 OPTIONS

RPO	Description	Qty	$ Retail
1FP87	Camaro Coupe	31,776	17,150.00
1FP67	Camaro Convertible	2,197	22,650.00
1FP87	Camaro Z28 Coupe	17,573	20,995.00
1FP67	Camaro Z28 Convertible	2,480	27,975.00
1LE	Special Performance Components Package	99	1,175.00
B4C	Special Service Package (same cost as Z28)	239	-
WU8	SS Package	3,025	3,500.00

1999 OPTIONS

RPO	Description	Qty	$ Retail
1FP87	Camaro Coupe	23,249	17,240.00
1FP67	Camaro Convertible	1,652	22,740.00
1FP87	Camaro Z28 Coupe	14,749	21,485.00
1FP67	Camaro Z28 Convertible	2,448	28,465.00
1LE	Special Performance Components Package	74	1,200.00
B4C	Special Service Package (same cost as Z28)	196	-
WU8	SS Package	4,829	3,700.00

2000 OPTIONS

RPO	Description	Qty	$ Retail
1FP87	Camaro Coupe	22,097	17,375.00
1FP67	Camaro Convertible	2,945	24,675.00
1FP87	Camaro Z28 Coupe	17,383	21,800.00
1FP67	Camaro Z28 Convertible	3,036	28,900.00
B4C	Special Service Package (same cost as Z28)	254	-
WU8	SS Package	8,913	3,950.00
Y2Y	SLP 2nd Sticker Content	5,159	0.00

2001 OPTIONS

RPO	Description	Qty	$ Retail
1FP87	Camaro Coupe	12,481	17,650.00
1FP67	Camaro Convertible	3,876	24,945.00
1FP87	Camaro Z28 Coupe	11,200	22,200.00
1FP67	Camaro Z28 Convertible	1,452	29,325.00
B4C	Special Service Package (based on Z28)	288	-
WU8	SS Package	6,332	3,950.00
Y2Y	SLP 2nd Sticker Content	5,373	0.00
Y3B	RS Package Through SLP	398	849.00

2002 OPTIONS

RPO	Description	Qty	$ Retail
1FP87	Camaro Coupe	14,221	18,455.00
1FP67	Camaro Convertible	2,750	26,450.00
1FP87	Z28 Coupe	19,472	22,870.00
1FP67	Camaro Z28 Convertible	5,333	29,965.00
B4C	Special Service Package	708	22,870.00
WU8	SS Package	11,191	3,625.00
Y2Y	SLP 2nd Sticker Content	10,625	-
Y3B	RS Package Through SLP	443	849.00

2010 OPTIONS

RPO	Description	Qty	$ Retail
1EE37	Camaro Coupe LS	7,767	23,530.00
1EE37	Camaro Coupe 1LT (27,430 w/ 2LT 1EH37)	51,469	24,730.00
1ES37	Camaro Coupe 1SS (34,500 w/ 2SS 1ET37)	70,169	31,795.00
B2E	Synergy Special Edition ($2,250 w/ automatic)	2,425	2,060.00
WRS	RS Package w/LT (1,500.00 w/ 2LT)	26,064	1,800.00
WRS	RS Package w/ SS	55,144	1,200.00
Z4Z	Indy 500 Pace Car Replica	294	7,155.00

2011 OPTIONS

RPO	Description	Qty	$ Retail
1EE37	Camaro Coupe LS	14,283	23,530.00
1EE37	Camaro Coupe 1LT ($28,075 w/2LT 1EH37)	22,940	24,730.00
1ES37	Camaro Coupe 1SS ($35,145 w/2SS 1ET37)	4,592	31,795.00
1EF67	Camaro Convertible 1LT ($33,500 w/2LT 1EH67)	2,580	30,000.00
1ES67	Camaro Convertible 1SS ($40,500 w/2SS 1ET67)	470	37,500.00
V9Z	Neiman-Marcus Edition	75	75,000.00

2012 OPTIONS

RPO	Description	Qty	$ Retail
1EN37	Camaro Coupe 2LS	13,411	24,480.00
NA	Camaro Coupe 1LT	18,442	25,280.00
NA	Camaro Coupe 1SS	2,497	32,280.00
NA	Camaro Convertible 1LT	4,054	30,180.00
NA	Camaro Convertible 1SS	370	38,330.00
NA	Camaro Coupe ZL1	1,971	54,095.00

2013 OPTIONS

RPO	Description	Qty	$ Retail
NA	Camaro Coupe 2LS	17,443	24,443.00
NA	Camaro Coupe 1LT	17,700	25,760.00
NA	Camaro Coupe 2SS	16,055	36,135.00
NA	Camaro Convertible 1SS	209	38,685.00
NA	Camaro Coupe ZL1	6,039	54,350.00
NA	Camaro Convertible ZL1	1,917	59,545.00

2014 OPTIONS

RPO	Description	Qty	$ Retail
NA	Camaro Coupe 2LS	20,952	24,755.00
NA	Camaro Convertible 1LT	5,474	31,005.00
NA	Camaro Coupe 1SS (w/1LE)	676	37,550.00
NA	Camaro Coupe 2SS	3,747	42,255.00
NA	Camaro Coupe ZL1	1,895	55,355.00
NA	Camaro Convertible ZL1	541	60,555.00

2015 OPTIONS

RPO	Description	Qty	$ Retail
NA	Camaro Coupe 1LT	25,147	26,005.00
NA	Camaro Convertible 2SS	3,857	42,405.00
NA	Camaro Coupe 1SS (w/1LE)	2,737	38,000.00
NA	Camaro Coupe Z28	1,292	72,305.00
NA	Camaro Coupe ZL1	1,561	55,505.00
NA	Camaro Convertible ZL1	289	60,705.00

2016 OPTIONS

RPO	Description	Qty	$ Retail
NA	Camaro Coupe 1LT	N/A	26,695.00
NA	Camaro Convertible 1LT (w/3.6 V-6 engine)	N/A	36,685.00
NA	Camaro Convertible 1LT (w/2.0l SIDI engine)	N/A	33,695.00
NA	Camaro Coupe 1SS	N/A	37,295.00
NA	Camaro Coupe 2SS	N/A	42,295.00
NA	Camaro Convertible 2SS	N/A	49,295.00

INDEX